W9-BFL-499

UNDERGROUND HOMES:
An Alternative Lifestyle

Other TAB books by the author:

No. 1172   *How To Build Your Own Underground Home*

No. 1372
$18.95

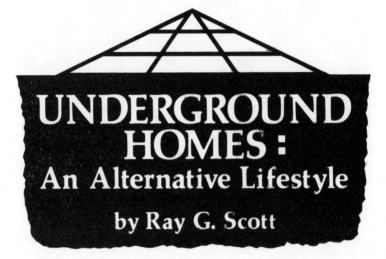

UNDERGROUND HOMES :
An Alternative Lifestyle
by Ray G. Scott

TAB TAB BOOKS Inc.
BLUE RIDGE SUMMIT, PA. 17214

FIRST EDITION

FIRST PRINTING

Copyright © 1981 by TAB BOOKS Inc.

Printed in the United States of America

Reproduction or publication of the content in any manner, without express permission of the publisher, is prohibited. No liability is assumed with respect to the use of the information herein.

Library of Congress Cataloging in Publication Data

Scott, Ray G
    Underground Homes :
    An Alternative Lifestyle
    Includes index.
        1. Earth sheltered houses—Design and construction.
I. Title.
TH4819.E27S35   643'.2   80-28693
ISBN 0-8306-9626-1
ISBN 0-8306-1372-2 (pbk.)

Contents

Preface

In my first book, *How To Build Your Own Underground Home* (TAB book No. 1172), I documented the construction of my home, to the best of my ability, just as it happened. Little did I realize that enough information would be generated to fill a second book after only two years of living underground. But it has, and this second book's appropriate title, *Underground Homes:An Alternative Lifestyle,* settles the question once and for all that building and living underground is a pliable alternative to conventional home building. Any book that is written, published and distributed is a result of considerable energy on the part of many individuals. For this reason, I want the following people to know I appreciate them very much—my children: Michele Renae, Vicki Lea, and Jon Christopher (for reasons too numerous to mention, during a traumatic year); my friend and banker, Bob Gentry; my friend and lawyer, Pat Sullivan; my friend and mother, Naomi Scott, whose efforts saved me through rough personal times; my special friend and advisor, Charlie Hodges, who is my link with real class and stability; and finally to my publisher, TAB BOOKS Inc., who helped a real novice get started.

Ray G. Scott

Chapter 1
Definition and Objectives

The definition of an underground house at first seems totally self-explanatory, but if you consider the limitless variations of terrain, weather and human nature, you can imagine the extremes that are possible. Therefore, a clarification, along with my definition, is in order here at the beginning of this book. You can imagine the extremes that might occur when somewhere, sometime, someone has set up a permanent residence in a cavern or cave, 100 feet or more below grade surface. On the other hand, there are homes, particularly in California, with approximately 4 inches of soil or sod growing on the roof only to act as an insulator against extreme hot and cold temperatures. These two conditions should definitely be considered the extremes of engineering ease and difficulty.

Therefore, to avoid covering problems and methods of construction that would probably never be encountered by a potential underground home builder reading this book, I have established my parameters of a typical (if there is such a thing) underground home to be 2 to 5 feet under the earth's surface. The biggest reason I have for suggesting 5 feet of earth over the roof of an underground home is that this is the point of best compromise. By compromise, I mean that 4 to 5 feet of earth gives you the most insulation for the least amount of weight. If you consider that earth, with an average amount of small rocks and dampness from rain, weighs around 100 pounds per cubic foot, it is easy to calculate how many cubic feet of dirt you will have over your head. Multiply the number of cubic feet by 90 pounds. This will give you an estimated total weight that your concrete roof slab will have to support. The more weight overhead, the more reinforced concrete you will need and the more it will cost.

According to my personal tests and calculations, 5 feet of earth will give approximately 90 per cent of the insulation value that 10 feet of earth will give you. However, the cost and strength of the roof slab to hold up 10 feet of earth would be unreasonable, probably three or four times more expensive than a slab of concrete capable of holding 5 feet of earth. Figure 1-1 is a bar graph estimating the per cent of insulating value in relation to depth of soil.

Just as the depth into the ground could be varied, the amount of vertical exterior wall surfaces covered by dirt is likewise varied. The previously mentioned sod-roofed homes have no exterior walls covered by dirt. All the walls are conventionally constructed. But the home which I have built and now live in has three and one-half vertical sides covered by a minimum of 4 feet of earth. In all fairness, I've seen a few good designs with only three walls covered by earth. So now having mentioned the extremes for exposed exterior wall surfaces, I'll make this suggestion: If you are really interested in building underground, go all the way. The engineering problems are basically the same regardless of whether you have one side, two sides or all four sides plus the roof covered by earth. However, the benefits are substantially increased in direct proportion to the more exterior vertical wall surfaces covered.

MENTAL AND PHYSICAL PREPARATION

If this definition doesn't scare you, consider one more thing. Before you make the final decision to build under-

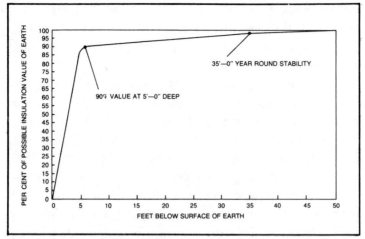

Fig. 1-1. Insulation value of the earth.

ground, you will want to be prepared mentally, as well as physically, to handle a project that is different from the normal. When I say physically and mentally, I mean exactly that. Building a house of any kind will test a person's physical stamina, especially when he tries to do most of the work himself. Remember that a concrete block soaked by rain can weigh nearly 100 pounds, that a shovel of wet dirt can weigh over 25 pounds, that a cubic foot of wet concrete can weigh 100 pounds a cubic foot, a sheet of plywood can weigh 80 pounds and that sheet rock might weigh 120 pounds. These are just a few examples of material weight.

So now you can just imagine the kick-in-the-head when some stranger (or friend) comes along and verbally tears your plans apart with inaccurate and incomplete knowledge or facts, after you have been working all weekend, or a distant relative drops in after a hard day's work and proceeds to tell you that you have a screw loose and probably means it. This is an example of the mental harassment you must be prepared to handle. Sometimes people can unintentionally be downright demoralizing. But then there's always a friend to come by and lift your spirits by telling you that you are doing a good job and how much he likes the idea of building underground. This is the type of friend you need. He is also most likely the same friend you can call on for a helping hand. The person who knocks your project probably won't hang around long enough to get involved or help out.

Contrary to popular belief, most people are not innovators or experimenters, and they don't know how to handle anyone who attempts to be one, except by criticizing. This is probably the truest statement you'll read in this book. In my short life of 38 years, I have so far done quite a few unconventional things; building an underground home is definitely the biggest risk from a financial viewpoint, but it probably won't be my last project. If I've learned one thing by my experiences, it is simply that the overwhelming majority of the population only wants to take a look at the unconventional. Many people will say they would like to do this and that, or someday they'll do such and such, but they really can't come up with a reason for not starting their dream projects immediately. They just procrastinate until it's too late in life to accomplish anything, whether it be an underground house, a sailboat, a trip or a job change. As you read this paragraph, I'm sure you will recognize yourself, friends and family. I hope you, the reader, are daring enough to be innovative whether you decide to build an underground house or not.

I am fortunate that most of my neighbors were and are kind people and seem to be sincere in their friendship and interest. But don't count on this attitude of acceptance of your endeavor. Expect to be called foolish, dumb and worse. If you are lucky and you have nice neighbors, things should go well. Consider yourself fortunate. But I will bet you my last dollar that there will be one joker in your neighborhood, just as there was in mine.

BENEFITS

Now that you know what I consider an underground house to be, and you have been warned of the mental harassments, let me now tell you of the benefits you can expect to find.

Fuel Savings

Since temperature, rainfall, winds and storms vary from coast to coast, some of your major objectives may be different from mine. The reason I decided to build my geothermic (sounds more technical than underground) home was definitely the cold winters and the hot summers of Maryland. In the winter the fuel bill in my previous conventional type home was doubling almost every year, and the electricity to run air conditioners in the summer wasn't doing any better.

Nature's Gifts

If you have ever taken a tour of any of the commercially-operated underground caverns, you'll remember seeing a sign somewhere near the beginning stating that the temperature at this point never varies from a specific degree, usually 54° or 55°F, winter or summer. This is one of the great gifts of nature which few of us take advantage of.

As you are probably aware, windmills and wind generators are enjoying a rejuvenated popularity since the fuel shortage of 1973. This is one way of taking advantage of nature's free gifts. Another, of course, is water power to turn a similar generating system. And don't forget that a few of the real back-to-earth folks (who deserve a great deal of credit) are using block ice frozen in the winter to keep cold storage areas cool through summer. As for solar power, I'll only mention it and suggest that you read up on the subject before building this underground house. There are millions of words written on solar energy. It's here now and forever, and it's practical to use.

These gifts of nature, along with many others, are used by only a small segment of the population because it's just not as convenient as they would like it to be. Geothermic heat and earth insulation are just as free, but only tested and used by a minute few. I feel, however, that this will be changing in the near future simply because the cost of all fuels will continue to climb at an unreasonable rate.

Once you are closed in 4 or 5 feet underground with average exterior exposure, you can expect to find year round temperatures stabilizing. Actually the temperature does not stabilize year-round until you are between 30 and 35 feet below the earth's surface, but the interesting fact is that after only 4 feet of dirt and 10 inches of concrete the lowest temperature I have recorded inside my home while under construction during the winter of 1977 was 46°F. This was without man-made heat of any kind. So instead of paying to heat your house in the winter from the mid teens to a comfortable temperature near 72°F, your additional heat requirement will be minimal. Remember, the constant temperature referred to under ground is only when there is no life activity. Once you add light bulbs, cooking heat, body heat and appliance heat, the temperature will be much higher and thus leave only a few degrees to be raised by conventional or experimental means. Also note that in the summer these same heat additions are not great enough to require air conditioning. The highest temperature recorded in our home during the summer of 1978 was 79°F. You must realize that every building and every location is as different as the people who build them, so you can expect some of these examples to be more in your favor, or less in your favor depending on your situation. I am only quoting from my own personal experience.

Maintenance

During the time my working design was forming, I nearly overlooked the other major advantages. Consider exterior maintenance. Since an underground home has no trim to paint, windows to wash, or shingles to blow off, and the exposed walls are stone, you do not have to put many hours into constant outside work. The only thing you really need is a good riding lawn mower and a small push power mower.

Permanency

As for a third reason, don't forget that all exterior walls are concrete, as are the ceiling and floor. So if they are designed and constructed correctly, they are impervious to nearly everything. Nothing—insects, fire or water—deteriorates the basic structure. So you can forget the annual termite inspection, rotting beams, etc.

Equally important is the elimination of major fire potential. If you build your house as I did mine, it is nearly impossible for a fire to get a foothold unless you are a pack rat and cram your storage areas with combustible items. Of course, even an underground home would be vulnerable to this type of carelessness.

Theft Factor

By eliminating all windows, you remove the temptation of vulnerable openings to a petty theft even though we know that the pro is going to find a way to steal regardless of the type of house you have. However, an underground house definitely gives you a feeling of security and stability.

For the Lady of the House

Remember when you don't have windows on every wall, you can arrange furniture in an endless combination because you never block a window. This should be particularly interesting to the woman of the household. Also, she will delight in the fact that she will not have to wash windows, nor will you have to buy as many curtains as in a conventional home.

Property Conservation

Last, but not least, is one of my favorite reasons. I bought approximately 2½ acres of rolling hillside in beautiful

Harford County, Maryland. After building a 3600 square foot house, including the garage, I still have 99 per cent of the acreage usable and unobstructed by any man-made objects. Besides, the children can't knock a baseball through my windows.

Now that you know all the good points and are thinking seriously about a subterranean home, as they are sometimes called as opposed to geothermic or underground, I'll tell you a few of the alternatives you can begin to think about.

ALTERNATIVES

The alternatives are unlimited—just let your imagination take over. For example, these houses can be circular, rectangular or square. There are one-story, split-level and probably two-story houses. They are as small as one or two rooms, or as big as my 40 foot × 90 foot, two-level house. Size and shape are only the first of many major decisions you will have to make.

Skylights or domes are usually necessary, sometimes covering indoor gardens, sometimes only providing light to rooms. You could have one dome or six skylights or anything in between. One underground house I know of has the main entrance by a staircase to the center of a circular layout. I consider this an unconventional underground house because of the irregular room shapes. Room locations and exits are your personal choice. The method of construction is what is critical. The most important thing to consider is that room location will have to meet building codes where applicable to your locale. I'll cover codes in a later chapter—read it carefully.

Another factor that enters into alternatives regarding your house design is the land or site you select to build into. Does your design fit the land? Remember, you can build an underground home almost anywhere except a swamp. Once the exact location is established you eliminate many of your alternatives. For example, if the land you decide on is rocky, you will not be excavating as deep as you would if it were a sandy soil with no rocks. And if it faces north instead of south, you certainly would not want to build the open side of glass; just as you shouldn't excavate very deep if standing water is close by. One fact you will notice as your plans begin to progress is that by the time you meet building codes, zoning regulations, neighborhood restrictions and avoid the natural pitfalls, you don't have the limitless possibilities you first pictured. Nevertheless, even after meeting all these regulations, an underground house will still be a challenge to the imagination. Keep tossing over in your mind all the things you could add to make your home interesting and individualistic.

Chapter 2
Evolution Of
Underground Building

The history books say that the Chinese lived in underground cities, the American Indians lived in cliff dwellings, and that the pioneers built sod homes; so the idea of underground structures has been documented as being a sound one for a few years. Yet the idea nearly died out as modern technology advanced. There weren't too many underground homes being built that I know of in this country until the price of crude oil started going sky high around 1973. Then all energy-saving ideas began to flourish; thus, underground homes came into the limelight.

AN UNFINISHED HOME

If you drop back and take a close look at recent history, you'll notice that even though people weren't building underground, they were flirting with the idea. Many of you may know of people who built a basement to live in while they finished the topside of their house, only to find 10 years later that the basement is as far as the house got. There are quite a few of these unfinished homes around even today. No one calls the home in Fig. 2-1 an underground home, but it's getting pretty close. In fact, it does share the benefits of

Fig. 2-1. With the roof exposed as it is, humidity becomes a problem.

Fig. 2-2. A one-car garage.

Fig. 2-3. View of a workshop.

heating and cooling to some degree with my house. The bad feature is that with the roof exposed as it is, moisture and humidity become a major problem.

OTHER BUILDINGS

Figures 2-2 and 2-3 show that using the earth as shelter has been accepted for outbuildings as well as human inhabitancy. These two photos are of a one-car garage and a small workshop. The insulation value, even in an open building such as in Fig. 2-2 is tremendous. The owner told me that at floor level in this garage, he has never seen it cold enough to freeze. This is quite an advantage for starting a car in the middle of a Maryland winter. Now I realize this isn't the greatest use of underground building around, but once again it is an indicator that the idea is commonly used by the average person, at least to some degree, and that the public in general is accepting the concept more every year. All this means to you, the prospective underground home builder, is that the roadblocks of financing, zoning and neighborhood approval are becoming easier to overcome.

Chapter 3
The Right To
Build Underground

This short chapter will briefly discuss a very important right that you, as an underground home owner/builder, have and may at some time have to fight to keep. It is simply your right to the sun.

YOUR RIGHT TO THE SUN'S RAYS

This problem of who can or can not block the rays of the sun or the view of the water by man-made interference, whether it be a building or intentional planting of trees, has been through the courts many times. The best I can figure is that the decisions are going about 50-50; that is, 50 percent of the time the party whose view is blocked wins, and the other 50 percent of the time the builder's right to build on his property is upheld. So no clear-cut precedent has been set, probably because each situation is so different. See Fig. 3-1.

This blocking of a natural view, or of the sun's rays, becomes very critical when living underground for a few reasons. First, most homes rely on some sort of passive solar action to heat their underground home. If the sun is blocked, there goes the heat. Second, since underground homes usually only have one wall exposed with windows,

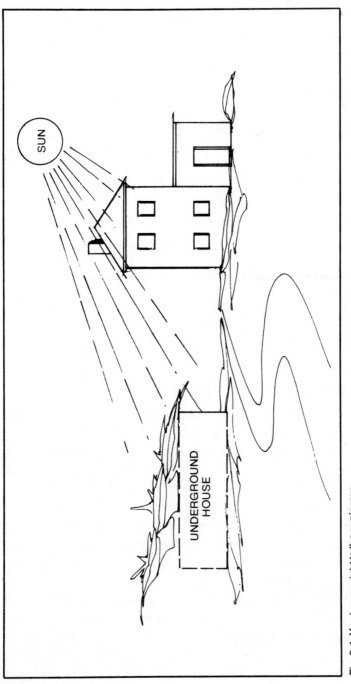

Fig. 3-1. You have a right to the sun's rays.

SUN

UNDERGROUND HOUSE

and a builder next door blocks the sun's rays away from those windows, it literally ruins all the plans for ventilation, heat or path of sight.

Third, an underground home is just that—very close to the ground. If you think about it, you will agree that it won't take a skyscraper to block the sun. In many cases, a single story home will do the trick.

BUILD ON A FEW ACRES

This is why I've mentioned in everything I've written about underground homes to get a few acres to build on. Then no one can block anything from you—sun, wind or view.

If by chance you end up with a small parcel of land, and construction begins next door, don't wait until the structure is complete to check it out. If you were there first, and you get a good lawyer, you might prevent a catastrophe that would leave you the proud owner of an underground white elephant. Remember, don't build underground unless you have at least two acres or more.

Chapter 4
What Are The
Real Advantages?

I have to believe that anyone contemplating building a house underground has done some basic research on the subject. By research, I mean reading magazine articles and, of course, my first book on the subject, which should be on everyone's "must read" list. (That's my personal opinion, of course.) Additional research could be done by contacting someone who has built an underground house and arranging for a tour. I have shown my house to hundreds of people who write or call asking for a tour. This first-hand look is absolutely the best way to get a real idea of whether you are interested or not. Many times couples have come to investigate. When they arrive, the man is usually the one who has the interest. The woman is skeptical, but looking. Within a few minutes the woman usually admits that her preconceived fears were not true. They usually leave convinced that the idea is feasible.

ADVANTAGE: HEATING

Turn to any magazine article on underground home building and the first paragraph will mention the biggest reason to build underground, which is to save heating fuel.

This is absolutely true, but the importance of this reason over the other reasons is out of proportion. So much has been written on the subject of underground homes not needing a heating system that everyone is beginning to believe it. I want to put that myth to bed once and for all. Every home regardless of type of construction, built north of the sun belt of the United States, needs a heating system. What these articles are trying to say is that it takes *less* to heat a properly built underground home than an average conventional home. The best estimate I can make without exaggerating to prove a point either way is that the actual dollar heating cost of an average underground home will be one-third that of an above-ground home (Fig. 4-1).

To say a house is being heated only by a wood stove isn't very exact. Any house, including the White House, could be heated by a wood stove if it were big enough and you had an endless amount of wood. To begin to put cost in true perspective, firewood is sold by the cord. A cord of firewood is a stack of wood approximately 4 feet wide x 4 feet high x 8 feet long. A cord of wood sells for approximately $85 delivered, as of January, 1980. The catch is that the cost of wood is going up faster than the price of oil.

The point everyone is trying to get across is that it definitely is economical to heat an underground house by wood. Of course, it's economical to heat one by electricity, also.

In summing up the heating advantage, it can be said that even though heating by wood seems to fit the atmosphere of an underground house, it probably could be heated just as economically by electricity. I'm not even going to discuss heating by fuels other than electricity or wood. It is my opinion that it is too dangerous to bring propane or a liquid fuel into an enclosed structure. I know some underground homes that have been built successfully against my opinion, but I'd rather be safe than sorry. Give my opinions serious thought. Don't use liquid or gaseous fuels for heating or

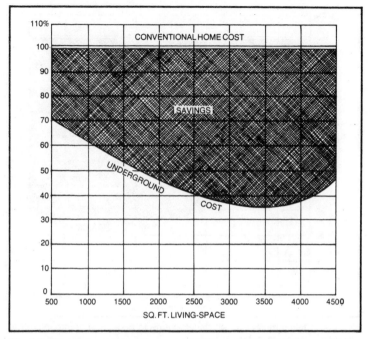

Fig. 4-1. The graph compares the cost of heating an underground home with the cost of heating a conventional home.

cooking. Build underground and save two-thirds on your annual fuel bill.

ADVANTAGE: COOLING

Since cooling costs are really electricity costs, you can only claim a saving by adding up the cost of the electricity you didn't use. The following example will illustrate what I mean. If owners of a conventional house nearby have an electric bill of $60 per month in the winter (they must heat by fuel oil), but the bill goes up to $90 in the summer, it doesn't take a genius to add up the dollars spent on electricity for air conditioning if both homes (underground and conventional) are holding the same temperature (approximately 74 degrees Fahrenheit). Figure 4-2 shows that it takes considerably less electrical energy to cool an underground home regardless of square footage.

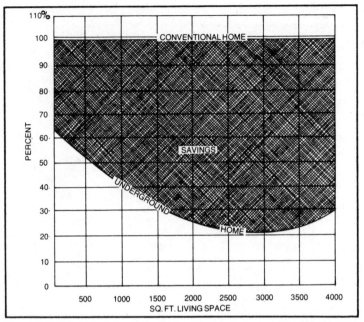

Fig. 4-2. It takes much less electrical energy to cool an underground home regardless of square footage.

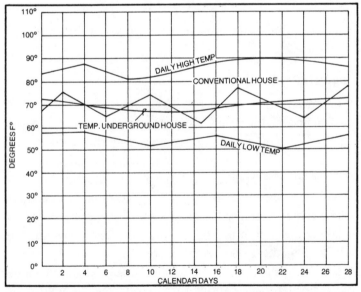

Fig. 4-3. The graph shows the temperature in degrees Fahrenheit inside similar houses with similar conditions.

Figure 4-3 shows the temperature in degrees Fahrenheit inside similar houses with similar conditions. The only difference is one house is underground and the other is conventional. Notice how an underground home fluctuates gradually, but a conventional home follows approximately a two day cycle.

Chapter 5
Insects

If you read my first book, you recall the discussion of bugs, insects and any other creatures that may try to inhabit an underground garden. I said that I didn't have an insect problem of any degree. In fact, the only uninvited living creatures that have shown up in any significant numbers (except humans) are crickets. Crickets are those black grasshopper-like insects that chirp at night. I assume there are a few sections of this country where crickets don't exist, so it is not unreasonable to assume someone has never seen one.

I didn't have a bug problem, only a slight cricket problem, and this needs clarification. That statement was made after the first year of occupancy. I thought the crickets were a nice addition as far as atmosphere goes, and they didn't hurt anything. Also, they stayed to themselves in the garden. All of these facts were true after the first year.

MULTIPLYING CRICKETS

Once springtime came in the second year, my cricket population came out in full force. The second spring and summer it was obvious that their population had increased, but that still didn't bother me too much. They still stayed to themselves in the garden. However, the nighttime sounds began to resemble the sound effects of old Tarzan movies. But once the heat of midsummer built up in this garden, the crickets sought refuge in the outside, probably under rocks or back into the woods or fields. So the second year ended with no major headaches from the cricket invasion.

DEALING WITH A CRICKET PROBLEM

The third springtime season came as scheduled and everything was as predicted. Trees, flowers, grass and people seem to react fairly predictably to the coming of spring, but not crickets. There must have been something romantic in the air for crickets this past winter, because at the first sight of warm weather in my garden out they came. So I resorted to inhuman tactics and used a can of bug spray. This slowed things down for a couple of days, and I thought the cricket invasion was over. I was wrong. They only retreated for fresh air and regrouped. About a week later the second phase of their invasion began. It got to the point where they chirped (if that's what crickets do) all day as well as all night. They began to challenge the radio for volume. By now all the elements for a great science fiction movie were beginning to unfold, so I thought I'd better call a halt to the invasion once and for all and establish my territorial rights one more time.

This time I obtained some commercial bug killer (like farmers use on their crops). I bombarded the enclosed garden and kept it closed off to human traffic for a couple of days. I think it's needless to say that the invasion was over. Now it was time to clean up the aftermath of cricket carcasses with a vacuum cleaner, and I did.

It's been six weeks now at this writing and the cricket population seems to be under control. I plan to do a controlled spraying every month or so to make sure they don't multiply out of hand as they did over the past three years.

CONSERVATION AGENT

My county has a conservation agent. He is a civil servant that is an expert in farm problems such as plants, growing, soil, fertilizer and, of course, insects. I checked with him and found that I had created a perfect incubator for the cricket-type insects, and what I was doing was operating a cricket farm inside my house. His advice and suggestions soon brought the problem under control, and we established a method that will keep it that way.

If you plan to build an indoor garden or have one already, I suggest you talk to this person on your local level. Sometimes he's called the farm agent, soil conservation agent or something similar. Every locale has their own name, and I think every locale has a person who does this type of work. By the way, this information is free to you because your tax dollars are paying for it. So use the agent's information to benefit you for a change; get some of your tax dollars back.

LACK OF OTHER BUGS

I can't understand why, but with this constant influx of crickets, other species of insects have failed to show up. Spiders, ants and similar creatures seem to not tread where crickets are.

Chapter 6
Security

Much has been written about the advantages of living underground. One of the advantages that always gets discussed after all the accepted major advantages are documented, such as heating and cooling, is *security*. And I admit to being guilty in my first underground home book of not giving security the proper coverage. The reason I didn't was because I didn't know, and only after two years of living underground have I learned to appreciate the feeling.

The security aspect that I think potential underground builders or buyers should be aware of is very real, even though it very seldom gets a chance to be proven. And only recently did I really become aware of how secure the general public views an underground home.

AIR POLLUTION

If you look on any map and locate the now famous Three Mile Island, Pennsylvania, and also find Forest Hill, Maryland, you'll see that they are about 40 miles apart, as the crow flies. Once the news of the Three Mile Island nuclear power plant disaster made headlines, I was asked many times by friends and neighbors if they could move into my underground house if the Three Mile Island power plant blew up. Of course, they were only joking at the time. It is easy to joke about something that hasn't happened yet. The point is everyone who discussed this with me had given it thought, and I'm sure they were convinced that my house would be a saving sanctuary.

I might as well put that idea to rest once and for all. Air is everywhere. What is outside your house will soon be inside, and only by a complex filter system can dangerous particles be filtered from contaminated air to make it safe for humans to breathe. If you are inclined to worry about nuclear power plant disasters, or chemical or biological warfare accidents, then you could develop a design that would incorporate an adequate filter system. I am not going into the subject of filters any deeper in this book because it is not pertinent to underground living.

One thing in your favor if you do decide to consider a clean air system is that an underground house lends itself very easily for adaptation. It would be very difficult to install any type of filter system in a conventional house because of the numerous places that air enters into the home such as windows, floors and even the roof.

As I mentioned, security comes in many modes, and some are very realistic to consider. Some aren't worth mentioning. I consider security from air pollution of any type a lost cause as far as what type of house you live in. An individual effort against air pollution, be it nuclear or smoke from a forest fire, is basically a wasted effort. It takes unlimited cooperation, regulations and funds to produce a change. Individuals have neither the time nor the money.

HIGH WINDS

The real security that I have come to appreciate (even though I haven't really had it demonstrated) is that against strong winds, especially *tornadoes*. Now I'm not telling you that Maryland is the tornado capital of the United States, but serious thunderstorms have been on the increase in recent years. Occasionally a tornado does touch down around here. I've never seen a tornado up close, but I can tell you the winds have come mighty close to being a real threat to the neighborhood. It didn't take long for me to catch on to the fact that I couldn't even hear the wind blowing once I was inside and the doors were closed. I would really hate to see anyone's home damaged by windstorms, but being a realist I know it's going to happen sooner or later. I can't imagine a windstorm strong enough to damage the grass on my roof.

FIRE

I don't care one bit that the building inspectors continue to insist that fire is a potential threat in an underground house. It just isn't there, at least in the underground homes that I've had the privilege to investigate. I will agree that noxious fumes caused by a burning chair or a similar incident could be a potential health hazard, but a full-fledged burning fire is very unlikely in my house. This is a feeling that I have grown more positive about after two years of having it brought to my attention.

THE CONCEPT OF SURVIVAL

There is a growing segment of our society that if categorized would be called "survivalists." These are a group of individuals who are gearing their lifestyle to survive a potential breakdown of society as it presently exists. Some of these people feel that a natural or man-made disaster could cause life to revert back to a more primitive time. Others think the economy will simply collapse because of too many give-away programs. Still others worry that war or racial violence will cause the downfall of our present society.

In summary, it matters not what the cause of this potential catastrophe would be, only that many people feel it will happen in the near future. If the subject of basic survival strikes an interest nerve in you, then I suggest that you should investigate it to the fullest. The library will be full of

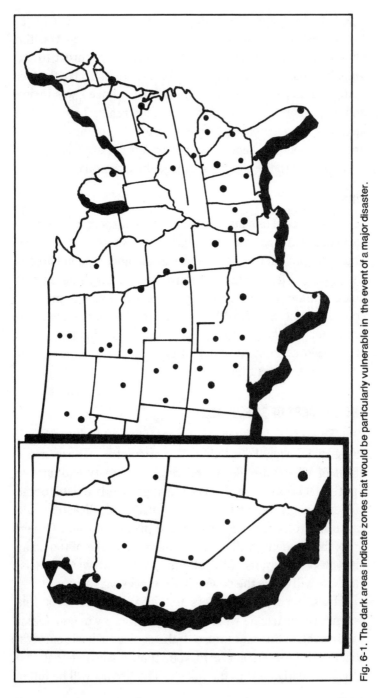

Fig. 6-1. The dark areas indicate zones that would be particularly vulnerable in the event of a major disaster.

articles, especially in the outdoor type magazines. Books exist on the subject, also.

You must understand that this is not one of my fears in life, and I haven't given too much thought to the idea of being a "survivalist." I personally believe that this country is the nearest thing to perpetual motion ever conceived, and the only thing that will ever stop it will be the theoretical end of the world.

The point I do want to make is very blunt. If anyone reading about underground homes is into this survival concept of future living, then you're on the right track. An underground home and the concept of survival go together like bread and butter.

If a home is designed like mine and a little thought is given to food and water storage, extra fuel and a generator, a "survivalist" could live literally years without ever coming out from underground. This would be quite a safe place to escape most every conceivable disaster, except for air contamination. I strongly suggest to anyone interested in joining the growing cult of "survivalists" that you start digging, the sooner the better.

Look at the map in Fig. 6-1. The dark areas indicate zones that would be particularly vulnerable in the event of any major breakdown of society. The lighter areas are more rural for the most part.

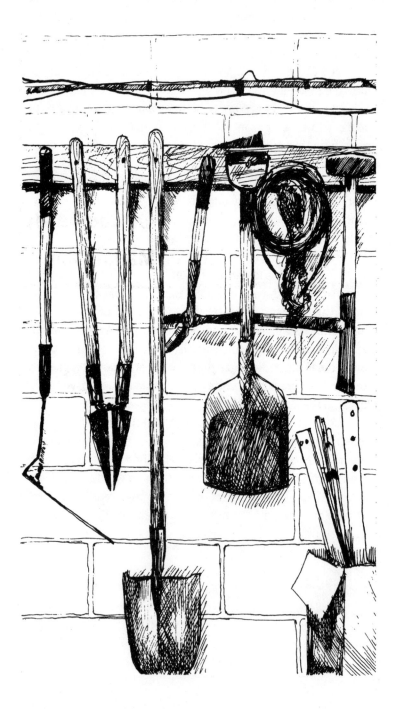

Chapter 7
Do You Really
Save On Maintenance?

Since the subject of maintenance, or lack of it, is always mentioned as one of the benefits of building underground, I figure a chapter ought to be devoted to a few facts and figures on the subject, based on three years of actual experience. Remember, this is first-hand information, not theoretical material that I've read in some magazine articles.

Maintenance is something that you control, whether you have a conventional home or an underground home. When I say you control it, I mean that the materials you choose to build with and the method of construction will have an effect on your maintenance time and money. I don't believe anyone will argue the point that poor workmanship leads to immediate maintenance. The same goes for cheap building materials.

ALUMINUM

While on the subject of building materials, there is no substitute for aluminum if you can use it. The cost is always much more expensive, but it's money well spent. The best example I can give is this—notice steel railings around patios or porches. It doesn't matter how well they are

Fig. 7-1. The aluminum railing will not rust.

painted or how often. Some spots will rust and drip down on the wood or concrete, staining it for good. The same railing can be bought in aluminum. See Fig. 7-1. An aluminum railing costs nearly four times that of steel, but I'll never have the problem shown in Fig. 7-2.

Today many items that were once only available in one material, such as asphalt shingles, are now available in aluminum with a wood-grain finish that looks like real wood, but it lasts forever. Aluminum is one of mankind's greatest

Fig. 7-2. Note the rust on the steel railing.

inventions. Use it to its fullest, and as an extra added bonus you'll get excellent insulation value, especially with the use of aluminum siding and aluminum roof shingles.

WASHING WINDOWS, LAWN CUTTING AND PAINTING

The idea of washing windows doesn't excite many people, especially me. I can definitely say that the absence of windows has been a blessing. If you don't have them, you can't wash them.

Keeping a lawn trimmed is a job with any house, but it is obvious that the size of the job increases with underground homes because of the additional retaining walls, landscaping, sidewalks, etc., generally associated with building underground, not to mention the area commonly referred to as the roof. It is now grass and needs mowing. So on the subject of lawn care, the fair thing to say is that the maintenance time increases proportionately to the size of the home (Fig. 7-3). It is definitely more work than a conventional home.

Needless to say, the less there is to paint, the less time and money it will cost to keep things neat and trim. So on this subject, there is also a definite reduction in maintenance time and cost.

DUST CONTROL

Many women have drawn the conclusion that it would be dust free in an underground home since there are no windows

Fig. 7-3. View of my lawn.

Fig. 7-4. This driveway can pose real problems in a snowstorm.

to open, but that's not true. If a home has as much carpeting in it as mine, the dust level is bound to increase. Most underground homes have carpeting throughout. You can expect the home to have a little less dust than a conventional home might have, but a big plus that causes a lot of dust is a wood stove. I use a wood stove as the sole source of heat and, believe me, it contributes its share of dust. This extra dust is a negative point as far as maintenance goes, but it is worth it.

DRIVEWAY

This is worth mentioning only to make you aware of a potential maintenance problem. Most underground homes

Fig. 7-5. A neighbor's driveway.

Fig. 7-6. My driveway.

are built into a hillside, thus requiring a steep driveway (Fig. 7-4). If you use gravel as a driveway base, it only takes a good, heavy downpour to put in a nice gully. Then you and your trusty shovel have to fill it in again. If you blacktop this driveway, the initial cost is high, plus your taxes go up. However, the biggest reason not to blacktop a driveway such as this is that in the winter it only takes a slight mist to freeze over the surface of a blacktop driveway (like my neighbor's in Fig. 7-5). It is impossible to get up or down with spreading sand. If you leave the driveway gravel like mine, a little snow or ice is never a problem (Fig. 7-6). Of course, this is true for all homes, underground or not.

SUMMARY

In all fairness to you, the potential underground homebuilder, I can safely say that you will gain no advantage worth mentioning on the subject of home maintenance. All homes are alike in the one respect that the effort to keep them neat and clean is going to be there whether built above or below ground. So forget the advantage of low maintenance.

Chapter 8
True Tests

Building a structure underground lends itself to three basic questions. First, will it hold up the weight? Second, will it leak? Third, will the heating and cooling benefits be realized?

STRENGTH AND LEAKS

The first question of strength was answered as soon as the construction was completed and it's still holding true over three years later. There is not one sign of structural failure.

The second question of leaks has been answered more gradually and is tested twice annually, once during the spring rains and once again with the fall rains. There is no problem here, either.

HEATING AND COOLING

The third question of heating and cooling benefits being a worthwhile reason for going to all the trouble of building underground takes a little longer to verify, especially if the seasonal extremes aren't severe. Well, three winters have passed and each one was a typical Maryland winter, a few cold days and a few cool days, and then the cycle is repeated. The summers are usually known to follow the same pattern, except for the summer of 1980. The state of Maryland, as well as much of the South and Southwest sections of the country, had a real heat wave—the first major heat wave since my house was built. The heat wave was the first important test of the cooling effect of building 5 feet underground (Fig. 8-1). For 12 days in mid-July, 1980, the temperature reached 98 degrees Fahrenheit or above. The majority of those 12 days reached triple digit temperatures.

This 12-day period was definitely the ultimate test for my design. The inside temperature of the home had leveled out at around 74 degrees Fahrenheit plus or minus 1 degree through the month of June, when the outside temperature had been bouncing around 85 degrees Fahrenheit. Then the temperature inside began to rise at the rate of ½ degree per day. You can easily see that when it is 74 degrees inside and you increase the temperature by ½ degree, you don't get concerned because you know that tomorrow will be cooler than today's 104 degree reading. But tomorrow comes and so does a temperature of 100 degrees, and the next day is the same. All of a sudden the inside temperature of your house is nearly 76 degrees and gradually climbing. Now concern begins to set in because the weatherman on television says that the end of the heat wave is nowhere in sight. To cut down on the day-by-day account, the heat wave lasted 12 days in Maryland. Inside my house the temperature rose to 80 degrees.

I do not know how high the interior temperature would have climbed had the heat wave continued, but the rate of increase per day had slowed down towards the end. All indications are that 80 to 81 degrees would have been the high had the heat wave continued for two more weeks. I doubt that I will get a chance to prove this theory because a month long heat wave of triple digit temperatures in the state of Maryland has never happened, and probably never will, in

Fig. 8-1. Front view of my home.

my lifetime. This is the most severe test I expect to witness regarding heat.

In summarizing heat penetration to the interior of my house, I'm convinced that a 19 to 21 degree differential from inside to outside would remain through any reasonable heat wave.

COLD WAVE

Cold penetration from the outside to the inside is a different ball game altogether. You must realize that what I'm about to write isn't practical to prove, since it would require moving out of the house and shutting off all utilities. Like I said before, it's not practical just to prove a point.

The point that I think I could prove would be that through a normal winter season here in Maryland, if you moved out of the house and shut off the utilities, the temperature would hold a much greater differential from exterior to interior. The best I can estimate is that 40 degrees Fahrenheit would be the greatest difference you would see. If it reached 0 degrees Fahrenheit outside for two or three weeks, I'm convinced that the temperature inside the house would never drop below 40 degrees Fahrenheit.

As soon as humans inhabit a dwelling, body heat is exhumed and lights are turned on. The hot water heater, washer and dryer are used. Much more heat is contributed, without intentionally adding heat. If you could eliminate using any heat from a heating system, including a wood stove, the house would never drop below 55 degrees Fahrenheit. Now that's a fact to think about. A home above ground would drop below the freezing point in approximately two days if the outside temperature were 0 degrees Fahrenheit and all heating systems were shut down. Needless to say, this would make life uncomfortable if not unbearable.

By stretching the point, I would never need a heating system of any kind, and I would survive. I didn't say live comfortably; I said survive. And at the rate the world is going downhill, that may be what I'll have to do someday. If things ever should get that bad, I figure my neighbors will be moving in with me. Then all that body heat would at least warm things up a few degrees. But I'm not going to sit up nights waiting for doomsday. It just isn't coming in my lifetime.

Chapter 9
Lawn Maintenance

Since it's summertime and the grass grows while you watch it here in Maryland, lawn maintenance is an important subject to discuss with potential underground home owners, especially if the design is similar to mine. See Fig. 9-1.

I designed these banks to be as steep as you see in Fig. 9-2. A lot of thought was given to how I was going to keep them cut. My ingenious idea was not original, but it was effective.

I've been cutting these banks every week in the summer for three years now with an old push rotary lawn mower. This was never a major job, because the lawn mower was lightweight and very easy to pull up and down with a rope (Fig. 9-3). Once I was cutting another section of my property with the same mower and tried to cut a small and well-camouflaged tree stump. Needless to say, the lawn mower came up the loser in the competition. The end result was a broken crankshaft and a discarded lawn mower. I borrowed my neighbor's rotary lawn mower to get through the week's bank cutting ceremony.

WEIGHT DIFFERENCE

I discovered immediately that all lawn mowers, like people, aren't created equal. There are probably 100

Fig. 9-1. Side view of my house.

different brands of mowers on the market today with every possible gadget or gimmick under the sun. The price range is from very reasonable ($85) to over $300.

Once I borrowed my neighbor's mower, I discovered what a hindrance excessive weight and gadgets were, and I began to appreciate my 8-year-old mower. As I started to shop around for a new one, I knew what I was looking for. Now in all fairness to the manufacturers of mowers, they are just another victim of government intervention with free enterprise. New federal regulations require certain shields for your safety, of course, but these all cost you money. You don't have any choice.

LAWN MOWER WITH A TWO-CYCLE ENGINE

Once you shop for a mower, you'll soon see what I mean by different styles. Now it's time to tell you which one I found to be the best by far. My old mower was a model marketed by Montgomery Ward. It was a three-horsepower mower and cut a path 18 inches wide. The model was the cheapest one sold by Montgomery Ward. Eight years later Montgomery Ward still offers this model. It is still the least expensive model they have and it weighs only 46 pounds (Fig. 9-4). This model has only the basic choke and manual wheel adjustment. It is an ideal piece of equipment for the job you

Fig. 9-2. Front view of my home.

might have. Also, it has a two-cycle engine. Why a two-cycle engine? That's a good question, but I've got a good answer. Four-cycle engines on most lawn mowers have a crankcase and separate oil supply just like a car engine. If you would use a four-cycle engine powered mower on a hill as steep as Fig. 9-2, it wouldn't take long for the engine to burn up because the oil would at times not splash around as the factory designed, because of the extreme angle. In case you don't remember, a four-cycle engine uses gas the same as a car

Fig. 9-3. Cutting grass is no problem.

engine. A two-cycle engine uses a gas-oil mixture and the oil is the lubricant for the engine. Therefore, there is no other oil supply. This way the engine is lubricated even if it is on a 60 degree slope. See Fig. 9-1. Need I say more about mowers, motors and underground lawn cutting?

MODIFICATIONS AND ROPE

Now you can make your own modifications to your mower if you wish. The modifications are removal of the chute deflector and the safety shield. I'm not telling you to remove these parts. I only said I removed them (Fig. 9-4). The reason is very simple. The safety shield will not allow you to pull the mower backwards up a hill with a rope, and the chute deflector gets caught on everything. Use your own judgment.

Once you have a suitable mower, you need a piece of rope to pull the mower up the hill. I found that a piece of ⅜-inch nylon rope about 25 feet long with a single knot tied every 30 inches or so and a snap tied to the end worked great. I drilled a ⅜-inch hole in the back of the mower body, snapped on the rope, and snapped it off when I finished cutting (Fig. 9-5). Be careful not to cut the rope. This is my method and it makes a potentially difficult job relatively

Fig. 9-4. Use the right lawn mower.

Fig. 9-5. This nylon rope is a useful addition.

easy. Shop around, find a system that works for you and be careful. Rotary lawn mowers are dangerous under any condition. I know from previous experience.

Once you have your mower fixed the way you want it, you'll notice that the handle has to be made to stay in a normal position because it flies forward going down a hill and upsets the mower (Fig. 9-3). Do not take the handle off or you can't use it from trim work elsewhere on your property. I never thought that an underground home book would talk about lawn mowers, but in my case the lawn mower is one of my most valuable tools.

Chapter 10
Ventilation and Air Exchange

This is a subject that causes as much concern to building inspectors and health department officials as any, especially when the topic of underground living is concerned. Most of their concern is unfounded. You must realize that the information in this chapter is going to be contradictory to any and all technical books you may have read. I doubt that many engineers or contractors will support my information, but nevertheless I'll tell you exactly the way the environmental conditions are in my house (3600 square feet). I will also mention the facts about air flow and ventilation that are commonly accepted in textbooks.

OXYGEN SUPPLY

There are a few variables that affect oxygen intake tremendously. First and foremost is how many living creatures (humans, pets, etc.) inhabit the building, and what percent of the time. If I remember my high school biology correctly, I think that humans need 18 to 22 percent (by volume) oxygen in the air they breathe to perform normal healthy activity. Humans could live for quite some time at an oxygen level of 13 or 14 percent. When oxygen in the air approaches less than 10 percent, it won't be long until someone is dead. You must realize that heavy smoking in the living area can screw up calculations on both the volume and quality of air required. This is critical if you are designing an air flow system relying on natural ventilation.

Back to oxygen intake. Most building codes require between two and three complete air exchanges per hour. If you question what an air exchange is, figure out the cubic feet of living space you have (not counting garage). It depends on who you are and where you are whether or not you include closets and storage areas, but figure length times width times height of all living space. This formula gives you the cubic feet of air in that area. These codes are dictating that the air it takes to fill that living area under normal pressure must be totally changed at least twice a day. The only thing I can say is that is nonsense. Everytime you exchange a cubic foot of inside air for a cubic foot of outside air, it is costing you energy. The cost to you is in the form of cooling, heating, dehumidifying or, at best, pumping the air in for the transfer of each cubic foot of air. Most calculations are based on the four-member family. Even if you don't have four members in your family, I agree that the air should be sufficient to safely sustain at least four people forever. Two to three times an hour is ridiculous.

OTHER OXYGEN USES

We all know it takes oxygen to support *combustion* or flames. The bigger the flame, the more oxygen that is required and the more air exchanging that will take place. No secrets here, right? Wrong! There is a secret of sorts, or maybe not so much of a secret but something overlooked by the individual home designer/builder, and it is important. If your wood stove is relying on oxygen (air) to continue burning, the faster it burns, the more air it draws out of your living area. This has to be or the fire will be smothered. However, the air that is flowing in front of your wood stove is heated interior air. So you are actually drawing warm air out of the room to feed the fire to go up the chimney in the draft. You can eliminate this, or at least control the percentage of air that is taken from the heated room, by adding a vent to the exterior (Fig. 10-1). This vent or duct will allow the stove to draw exterior unheated air to feed the fire, leaving the warm air intact in the room. This duct can be installed or run in a million different ways. Whatever you do, don't use plastic or PVC pipe. Outside of plastic pipe, any metal piping could be used. Discuss this idea with someone who sells wood stoves and knows their business. You realize I can't suggest a system to use since every installation and situation is different, but the idea is one to consider.

FRESH AIR DUCT TO
GARAGE OR
STORAGE ROOM
3" METAL DUCT

Fig. 10-1. Fresh air is vented to the wood stove.

COOKING ODOR

Cooking odors are another reason air requires exchanging. Cooking definitely causes an odor in the air (usually good). However, writers of buildings codes don't necessarily care if the cooking odor is pleasant or not. They want that odor gone. This is where a great deal of fresh air would be required to dilute the odor of a home-cooked meal. I agree that it would take constant air exchange to accomplish this. My point is that in this day of energy saving and non-waste, it is time to put up with a little cooking odor in the air rather than exchange that air for fresh exterior air. You never get something for nothing. If you want to get rid of all cooking odor, it is going to cost you. It is very simple.

It's time to remind you again that the information enclosed in this chapter specifically, and in this book in general, is based on data I have compiled from my experiences of living underground for nearly three years. I never intend to lead anyone to believe that my methods and data comply with any building code or, for that matter, any other professional in my field of expertise. I am telling you that the data and methods covered in this book have worked very well for me and my situation. The fact that my methods may violate one or more codes is a problem I faced and handled legally. You could do likewise or you could totally comply with codes as they are. This is your decision and only yours. Don't blame me if you follow codes without question and your house ends up costing 10 times what mine cost, because I didn't follow the codes too closely. You're not comparing apples to apples. You're comparing apples to oranges, and it doesn't work.

NEEDED AIR CHANGES

Here comes a real kicker as far as surprise information goes, so you had better reread the first part of this chapter again. I do not have a duct system to supply fresh air from the outside. I have no fan drawing fresh air in from the outside. I rely on *drafts* for my total air exchange. By draft, I mean natural leakage of air under doors, while they are shut, and through doors as they are opened and closed by routine traffic.

I tried to figure exactly how many air exchanges I actually incur in a day, but the air flow pattern in any house under normal living conditions is too sporadic to measure accurately. The information I was able to compile indicates that my house actually has a complete air exchange (approximately 20,000 cubic feet) every couple of days, as opposed to nearly three times an hour as required by code. If you do some quick figuring, you will see that my air exchange is approximately one-sixth that required by code. To look at it in terms that everyone understands clearly, I have to heat or cool, depending on seasons of course, one-sixth the volume of air that a conventional house owner would need to. I'm not saying that your heat bill would be one-sixth that of the heat bill for conventional housing. Even though the air turnover isn't nearly as great, there is a certain amount of heat absorption into the surrounding earth. I'm confident that if I were heating by conventional fuel with a conventional heating plant and duct system, the actual fuel bill would be much less than that of a comparable conventional house.

Chapter 11
How Underground
Homes Affect a Neighborhood

Rain runoff is a subject that seems to have been ignored by most of the articles I have read on underground homes, and it could be a legal problem if you are not careful. I'm going to cover the subject in this chapter about neighborhood relations rather than the ones on problems because I think it is more of a neighborhood acceptance situation than a problem. Of course, if you are sued for property damages, you could have a real legal problem.

RAIN RUNOFF

The term "runoff" refers to the direction and velocity that natural rain water (or melting snow) flows from your property. The direction that it flows naturally is the accepted course, and that's not your problem. What you are responsible for is any water that you divert onto your neighbor's property by rearranging the terrain of the property yourself. This is not usually a cause for concern with a conventional house. By building an underground house, you break a lot more ground and do much more grading than with conventional home building.

The only neighbor that I had who was unreasonable tried to make a big deal out of the rain water runoff onto his lawn. He said it devalued his property. I was threatened with legal action, but he finally moved out of town and that put an end to it.

Clear rain water isn't really going to cause anyone a problem unless it is in fantastic amounts. Rain water will remain clear and clean if your lawn has a good grass covering. The biggest problem I can visualize anyone having is in the initial stages of building before grass gets a good start. The accepted way to prevent water and sediment damage to your neighbors' properties is by stacking bales of hay along the property line. This method is required by building codes in some instances. Even if it is not, it is a good idea and it shows an intent to protect your neighbors' rights.

AESTHETIC VALUE

This is where the relationship between neighborhood and underground house can cause serious conflict and misunderstanding. I want to establish a fact right now that will explain my position on this subject a little more fully.

I built an underground house in a conventional, existing housing development in a moderate suburban Maryland community. This act could be put on the same parallel as the first black family moving into a previously all-white neighborhood in the mid-1960s. It didn't really hurt anybody, but it got everyone's attention and concern.

The reason I bring this to your attention at all is that even though legally there isn't much anyone can do to stop you from building your underground home (if you meet all the codes, zoning, etc.), it is much nicer to have a friendly relationship with your neighbors. Since I'm writing mostly about my situation (that is, having neighbors close by), I had to do all the public relations work possible with everyone in sight to insure a satisfactory relationship. Of course, if they aren't receptive or reasonable, do whatever you have to do. But don't knuckle under just because somebody doesn't like your idea. If you like it, that's what counts since you're paying the mortgage. I don't think there is another underground house in a conventional housing development anywhere in the United States. If any knows of one, tell me about it.

LOW KEY APPROACH

As I began building, I tried to keep a low profile of visibility. This was my intent. However, this somehow became impossible due to the newspaper coverage and building-inspectors harassing me. I still think a low key approach to anything new is the best way to handle it. If you can't be low key, at least be friendly and courteous.

TRAFFIC INCREASE

For the first two years that my house was under construction, the natural curiosity of people created more traffic in my development. I'm sure it irritated some neighbors, but I couldn't control it. Now two years later the extra traffic has dwindled to normal.

REAL VALUE

The argument that an underground home can or can not affect land value in a specific area isn't worth the breath it takes to discuss it. I heard the argument from the day I started to the day I finished building. I began to think I might have really created a condition that affected my neighbors' property value, until the neighbor who was doing all the yelling decided to sell his house. It was immediately next door to my property. His house went on the real estate market place at a selling price exceeding the neighborhood average and sold within a month or so. There goes the theory down the drain. Underground homes do nothing to affect real estate value one way or the other, assuming that your design is pleasant and the work is done professionally. I rest my case.

Chapter 12
Interior Furnishings

If these photos look familiar, you must have read my first book on underground homes. *How To Build Your Own Underground Home* (Tab Book No. 1172). It seemed unnecessary to change my interior furnishings just because they were featured in a book. Besides, we have added a lot of decorative improvements since the last book.

FAMILY ROOM

See Figs. 12-1 through 12-3 for close-ups of my family room. This room gets most of the traffic from three children and their friends, but the furnishings were bought to be durable. Figures 12-4 and 12-5 are of the dining area, also a heavily traveled area.

KITCHEN

Figures 12-6 and 12-7 are of the kitchen area. Notice the modern, all-electric appliances. (Don't notice the decorative wine bottles over the cabinets; we found them.)

BEDROOMS AND BATHROOM

Figure 12-8 is one daughter's room. Her personality shows in her choice of design and colors. Figure 12-9 is my

Fig. 12-1. Close-up of the family room.

second daughter's bedroom and, likewise, shows her personality traits. Figure 12-10 is my only son's room.

What can you say about a bathroom? The tub is separated by a partition from the dressing area. This is the children's bath and the largest. See Fig. 12-11.

Fig. 12-2. Different view of the family room.

Fig. 12-3. The most used room in the house.

Figure 12-12 is the main hallway leading from the family room to the bedrooms and through to the garage. Note the antique barn wood; it is over 100 years old, as is the brick behind the wood stove. This concludes the tour of the children's side of the house.

Fig. 12-4. The dining room.

Fig. 12-5. One of my favorite rooms.

Fig. 12-6. My modern kitchen.

Fig. 12-7. The kitchen is totally electric.

Continuing through the garden, you end up on the adult's side of the house. First you enter my bedroom. See Figs. 12-13 and 12-14. Note the walk-in closet. Through the door in Fig. 12-14 is my bathroom. Then we move into my

Fig. 12-8. The mural brightens the room.

Fig. 12-9. Personality shows in room decorations.

Fig. 12-10. Note the picture showing the view of the earth from the moon.

83

Fig. 12-11. One of two bathrooms.

home office where most of this writing was done. See Fig. 12-15.

OTHER AREAS OF THE HOME

Through the last door you come to the formal living room (in name only). The children have taken it over as a

Fig. 12-12. View of the hallway.

Fig. 12-13. A "one of a kind" mural design.

Fig. 12-14. As you can probably tell, I'm a bachelor.

Fig. 12-15. My personal hideaway.

music equipment and practice room. This accounts for the
lack of furniture (no room). This concludes the tour of living
areas, but that's not all of the inside. If you go back to the
family room and make a left, you come to the coat room that
doubles as a sewing room. See Fig. 12-16. Behind the coat

Fig. 12-16. The coat room doubles as a sewing room.

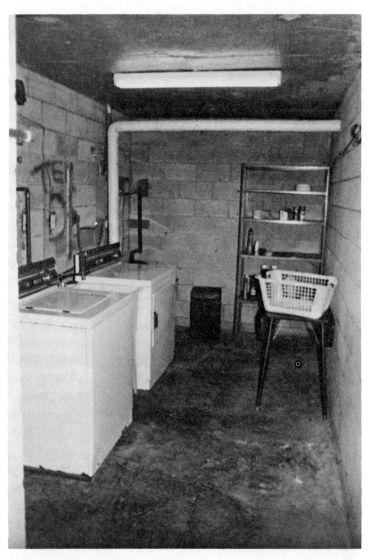

Fig. 12-17. The laundry center.

room is the very important laundry center (Fig. 12-17). Once again, notice the all-electric appliances.

Finally, Fig. 12-18 is of the storage area which is right next to the main entrance foyer (notice the brickwork). If you pass through the utility room (Fig. 12-19), you come to the

Fig. 12-18. A storage area.

Fig. 12-19. You never have enough storage room.

four-car garage. If there is one thing interesting about my garage, it is the 12-foot high ceiling. It was made that high to clear a large motor home or other recreational vehicle. Twelve foot high ceilings have many uses, to which any handyman will attest.

Chapter 13
Designing An
Underground House

The key phrase in this subject is "know what you are doing."
If you are not a professional engineer or a heavily experienced building contractor, don't go off and get yourself into trouble that could cause you financial disaster. Just because a friend of yours is a great electrical engineer or a high-paid civil engineer doesn't mean he knows a thing about strength of concrete, pouring methods, moisture problems, and all the other specific details he would need to know to advise you safely.

On the same note, don't rely on a builder of conventional homes to become your resident expert on building underground. Be reasonable. What does the strength of 2 × 4s and plywood above ground have to do with concrete under 50,000 pounds of earth? Absolutely nothing. It would be like getting a bicycle repairman to rebuild the engine in your Lincoln Continental. He would be basically mechanically inclined and might luck out and get all the parts back in the right order, with everything working fine. He may take a lot of trial and error work, or he may never get it right. The point is you will pay for his mistakes, and if they show up immediately you will be lucky. The mistakes an incompetent engineer or contractor could make might not show up until a couple of years later. They have been paid off. You own all the problems now. The cost to repair these problems would be tremendous because of the excavation, assuming that the problem could be corrected. Once again, make sure you are going on sound advice with competent people, or you will pay the price of shortcuts.

This is also a good time to insert a few words of wisdom that I have proven true time and time again. Someone else wrote them originally, but I proved them true. They are, "There is never time to do things right the first time, but always time to do them over." And the other is, "If something can go wrong, it will."

AMATEUR DESIGNING

There is a common "play on words" that is used daily in the English language, and most people never realize it. I think it should be clarified. I am sure someone will disagree with my analysis, but it's my book. The word I am talking about is "design." If you sit down with a pencil and sketch pad and draw a picture of an underground house, you most likely would tell your friends that you designed your house (Fig. 13-1). You feel you did, because in high school that's the terminology that was used. In reality, you drew an artist's conception of your house. Whatever the product is, even if you showed doors, windows, domes, driveway, etc., that's still an artist's conception. It is necessary, but not real designing.

PROFESSIONAL DESIGNING

Real designing is done by someone who can tell you how thick, what size, what material, strength, and so on that steel should be to build a bridge to hold 1,000,000 pounds of automobiles at one time in an 80 mile per hour windstorm, when the temperature is 10 degrees below zero and the bridge will be in use for 100 years. That's the work of a designer. The same point should be taken when planning

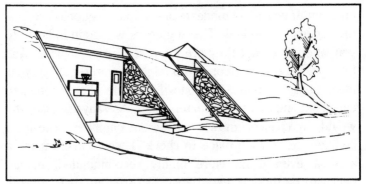

Fig. 13-1. An artist's conception of an underground house.

your underground house. If it's not designed to withstand the stress and strain of weight, time and the elements, you are in trouble. A picture doesn't begin to cover those little details.

WORKING WITH AN ENGINEER-DESIGNER

The best way to come to a happy conclusion if you are really serious about earth-sheltered living is to find an engineer-designer and work with him on the design. If he is not interested in your project, move on to another source. Nothing will cause more problems than a man designing something he couldn't care less about, plus it will cost you more.

If you haven't read my first book I will touch briefly on how to get an engineer to work with you. They are only human and respond to the same approach as anyone else would on any other subject. If you ask the right person the right question, you will get the right answer. If you live in a metropolitan area, just look in the yellow pages under "engineer-architectural" and you will find them listed proportionately to the population of your area—with lots of engineers in heavily populated areas.

Call each one and tell them your plans to build underground. Indicate that you are a private individual, not a contractor, and that your funds are limited. If they are not limited, just pick the engineer with the best reputation and hire him. If you are of modest means, like most of us are, use my suggested method. The engineers will soon indicate to you whether or not they are interested in your plans. Ask them to suggest someone who might be interested in your idea, if they are not. Soon you will compile a list of names to check out and talk to in greater detail. Ask anyone else who would, by chance, know a competent engineer. The local colleges are a good place to check. I would avoid using a student even if he were highly recommended, mainly because on-the-job experience is greater than all the textbook information in the world. An honor engineering student that has never waded in concrete yet isn't house-trained. Let someone else provide his on-the-job training program.

Another good source would be big construction companies. Many times the owners or managers are great engineers or at least travel in social circles where good engineers are known to exist.

If you live in a remote section of the country where the yellow pages of the phone book are minimal, you have a problem. Just because you are in the hills doesn't mean that you can overlook design. I've talked to people who said they didn't need an engineer because their locale didn't have building codes. I think that it is great not to have codes, but the logic of these people could kill them, especially if the amateur-designed concrete roof slab gave way. Don't confuse adherence to building codes with structural safety. There is little if any connection. I'll cover building codes in another chapter.

Once you find an engineer-designer you are comfortable with, sit down and show him your sketches. Explain any and all ideas you have and want to incorporate. Ask his opinion. He'll explain why this and that will cost more than it's worth or suggest a better approach. Ask what his charge will be, what he will provide and how much of the legal legwork he will do. By legal legwork I mean permits, investigations, appeals surveys, etc. You'll soon feel comfortable with your choice, if he is the right man. His charge will be directly related to his input from a time standpoint. The more rural your location, the greater the cost.

A word to the wise would be to check closely if you decide to buy a set of drawings already packaged and for sale by a mail order firm, as advertised in the "back to nature" magazines. The drawings are probably great, accurate and detailed, but there is the problem of adapting the pre-designed house to a completely different type of land base and terrain. If you are really set on going with a pre-designed drawing package, just hire an appropriate engineer to adapt it to your landscape.

As you and your engineer are working out a design, don't feel that you can't express your likes and dislikes. Room layout, for example, isn't critical unless you are trying to build an underground ballroom. The room layout is the principal part of the design that a homeowner usually wants to put his seal of approval on, and that is a minor part of the design problem. The actual location of a closet or bathroom usually doesn't affect any design information.

The location of skylights, stairways and garage doors are things that are controlled. So my suggestion is go at the design seriously. If you don't enjoy doing it and aren't happy with the design as construction starts, you most likely will have second thoughts, mostly negative, as the house is built.

LIVING WITH YOUR DESIGN

This could be one of the most important sections for an amateur designer to read and comprehend fully. Read it carefully and then again. You do not go around making arbitrary changes like moving walls and doorways once your underground house is built. First, it is difficult and messy to break away concrete. Second, and most important, is that it could affect the structural integrity of the total building. Of course, the cost of replacing walls isn't cheap.

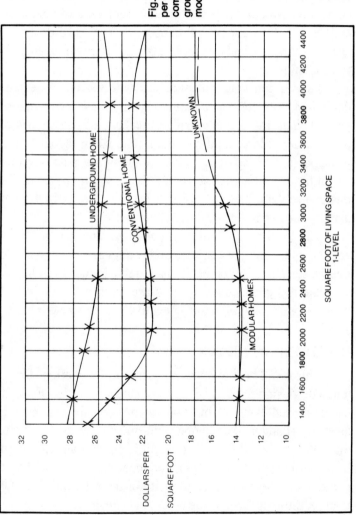

Fig. 13-2. The graph shows the price per square foot of living space and compares conventional and underground homes to the prefabricated modular ones.

If you have a conventional house and you want to change a doorway around, it is a simple matter of Sheetrock, 2 × 4s and a little labor. You can even remove a complete wall in a normal house with minimal trouble and expense. So it's no earth-shattering decision to make a major change even if the house is in final stages of construction, as long as it is above ground. If you try a trick like that once construction starts underground, you had better get your checkbook reinforced because it's going to cost you time and money.

All I am trying to emphasize is that you should be happy with and sure of your design before construction starts. It will become a monument of either your good judgment or bad judgment. Once built, there is little you can do to easily change things, but your payments will go on and on. While on the subject of living with your design, check Fig. 13-2 to see what happens to the price per square foot of living space. The graph compares conventional and underground homes to the prefabricated modular homes, the ones that come to the site on wheels. You can see that the construction of an underground home is the most expensive way to go, but it is worth every penny.

Chapter 14
Design and Land
Must Work Together

The design of your house and the contour of the land must work together. This is true regardless of what type of home you are building, but never more true than when designing an underground house. It's a fact that any land with a water table low enough to build a conventional home could be used to build an underground home, but a water table varies from site to site. The water table is the depth underground at which you first come in contact with standing water under normal prevailing conditions. It is also a fact that the drier the land, the easier your underground endeavor. Whatever you do, don't buy the first piece of ground you come across. Look around until you have found a couple of acres you really like. If there is such a thing as the ideal location for an underground house, it would be the top of a knoll or hill in the high section of your locale. It is also obvious that you can't always get the best, but since this is one of the most important decisions in your life, make it the best possible. Look very carefully at the drainage adjacent to your potential house location. Pretend that it has been raining for a week and imagine where the run-off would be going. This is the first step to judging the usability of the land.

WELL DEPTHS

Next, check the depth of some of the wells drilled in recent years adjacent to your choice of property. You will be looking for places where well depths are 200 feet or deeper. This tells you that water is not near the surface of the land you will be building on. This is not to say that if the neighbors have a well only 100 feet deep that you can't build an underground home nearby—it's only an indication that moisture is closer to the surface, thus nearer your potential house.

EXCAVATING

The type of problem you could run into is this. If by chance, you started excavating in mid-summer when most sections of the country are entering their driest months, you could be deceived into thinking the land is really ideal, then when the following spring rains begin (as they always do), the land around could turn into a big swamp, holding water like a sponge. If this would happen, of course water would try to come through the concrete wall, floor or roof.

WATER PROBLEMS

The action of water coming through a concrete floor of any house is hydrostatic pressure. Briefly, this is when the soil around the concrete block cannot absorb any more water and the excess water cannot run off because it is surrounded by a less porous-type soil or rock formation. At this point, the water has nowhere to go but through the path of least resistance, usually your concrete floor or block wall. This is a condition you must avoid at all costs when locating your underground house. If possible, wait until you have a rainy period. The greater the rainfall, the better. A positive way to see what the conditions would be around your house if it were already built would be to dig a hole 20 to 25 feet deep. It should be big enough to climb down into, but be careful of caving walls. Examine the soil close—under actual conditions. If you don't have time to do all this investigation, or it is not feasible for some reason or you would prefer to have a professional give you advice, look in the yellow pages of the telephone book under engineers—soil testing or engineering—soil. Every community of a reasonable size will have one or more listed. Anywhere there is major construction going on, especially road work, there will be soil testing facilities close by. This will cost you, but it is definitely worth the expense if the land you're thinking of buying is questionable. Remember, water leaking in your underground house could make it useless, and you would lose everything that you have invested.

COUNTY REGULATIONS

Last, but not least, your county probably has a department called *Land Use* or some department with a similar title. These local offices are usually a good source to check once you have limited your search for the land to one or two parcels. They have topography maps in great detail concerning the lay of the lands around your proposed home site. Also they have soil testing performed by Farm Bureaus for crop growth. All this information will be helpful to the person trying to make up his mind about a piece of ground. Also, the Health Department in most locales could be helpful, depending on their function within the local government. Be prepared to meet resistance or limited cooperation if these local officials know you are contemplating building underground. My suggestion to you, at this stage, would be to keep your plans to yourself. Remember, this is the voice of experience talking. Use all these sources to their fullest, combine them with good judgment and you shouldn't go wrong.

I strongly suggest that you make the final decision on exact location and get the deed in your name before beginning the design of the building. The reason I say this is that many things can happen on the way to the lawyer's office for final settlement.

FINANCING

You may not get financing at a reasonable rate, or you may not get financing at all. Let me tell you what happened to me because it may very well happen to you and you can plan accordingly. Since I'm building my house in the community that I was born and lived in all my life, as did my parents and relatives before me, you can see that the local banks were definitely on a first name basis with me. Aside from my life-long residence, my credit was flawless. When I decided to build a new house (my underground intentions as of yet unannounced), I stopped by to see my friendly banker. After a short discussion on houses in general, he asked me how much I needed. I gave him my figure and he said no problem. Within a few days a letter came in the mail stating that I could have the amount I requested—just stop in when I was ready to finalize the loan. It was just that easy, even though I was going to be my own contractor. This fact may bother some banks, especially if you don't have the credentials to back up intentions of being your own contractor. But this didn't bother my bank. Remember now, to this point they know

Fig. 14-1. A scale model of the outside of an underground home.

Fig. 14-2. A scale model of the inside of an underground home.

nothing of an underground house. So I have the letter of loan approval, but as I am a basically honest individual, I decided to tell the bank of my plans to build underground. As a matter of fact, I even built a scale model of my house (Figs. 14-1 and 14-2) to impress the vice-president. He was impressed all right—so impressed that he said he couldn't possibly approve a loan for a far-out venture like an underground house.

At this point the negotiations began and after a great deal of pleading and promising my life away, they changed their minds and approved my loan. But this was only because of my life-long ties to this particular bank. If a stranger or a younger, less proven individual approached my bank or any other loaning institution for money to build underground, I'm afraid the answer would be short and sweet. No, No, No. This story is not meant to discourage you, but only to emphasize what you are up against mortgage-wise. As for V.H.A. and F.H.A. loans, they are even more difficult to obtain. If you qualify, check into them, but don't waste time begging or waiting for government assistance.

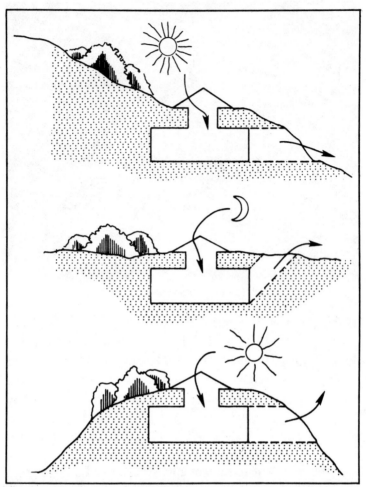

Fig. 14-3. Three basic possibilities for location of an underground home.

Another problem could be that the restrictions may not allow an underground house, or the soil may be unsuitable. These are just a few of the potential roadblocks to actually getting a deed in your name. If you invest your valuable time and money designing a building for a particular site and the deal falls through, it would be unfortunate and costly. If a professional draws up your blueprints, he will adapt to a particular piece of ground. If you change locations, the prints would have to be revised to suit the new location. This is not necessarily true for a conventional house, but most likely for an underground house.

LOT SIZE

I also suggest that the size of your lot be no smaller than two acres. The reason for this statement is purely cosmetic. Conventional homes can be designed to be attractive side by side on small plots of ground, such as many developments are, but in my opinion, an underground home loses much of its appeal when crowded by conventional homes.

Now let's assume that your dream location is a reality. The deed is in your name. If you have picked out a good piece of land, 50 percent of your potential problems will be eliminated. See Fig. 14-3 and locate the lay of the land that most resembles yours. Note the house location in each situation as I will explain the pros and cons of each as I see it.

INTO-THE-HILL

Into-the-hill is the most popular approach to building underground (Fig. 14-4). This into-the-hill method is by far the easiest to build, especially from a grading viewpoint, because approximately 65 percent of the cubic footage has to be excavated. The reason this is ideal is that you will need

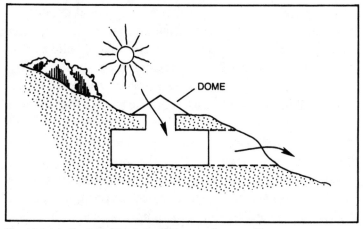

Fig. 14-4. Into-the-hill method of building an underground home.

tons and tons of dirt to backfill and complete final grading. This into-the-hill method provides a natural method of moving building materials around by an ordinary truck, because the upper part of the hill allows you to work at roof level. At the same time, the natural slope will give you access to the lower level by normal vehicles. If it were not for the ability to deliver building material to the lower level by truck, you would have to keep a crane of some type on the site to lower the heavier building material, and those cranes are expensive to rent. I suggest, for this reason, above all others, that you try to build the style of the underground house shown in Fig. 14-4.

LEVEL GROUND

The *level-ground method* is to be used as a second choice for quite a few reasons (Fig. 14-5). First, backfilling is much more expensive and difficult because the structure is above ground and since you did not excavate, you have to obtain fill dirt from somewhere to cover it with. This is definitely a course to take as a last resort. I had to buy some extra dirt for my final grading. I called every possible source to check prices of a dumptruck load (approximately 14 tons) and I found the prices (in 1978) ranged from $3 per ton to $10 per ton for the best quality top soil. So you see, buying the good earth isn't a cheap approach. Even if you could get dirt for free, you would have to rent a crane and bucket to put the dirt on top, because it would be unwise to take a bulldozer on your top slab, regardless of its designed strength. This would be expensive and time consuming. The extra equipment combined with the cost of buying and hauling soil will make the style more expensive than you probably want to get involved with. A second point to consider when looking at the ground-level style is that the heat loss and heat absorption is greater when a mass of earth is less in volume and above the natural terrain. There is one positive aspect of this type. It may be easier to comply with local building codes, simply because you would have the ability to exit from

110

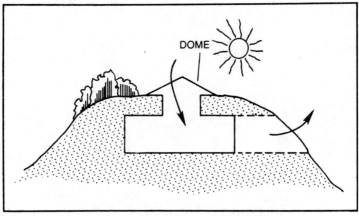

Fig. 14-5. The level-ground approach to an underground home.

each exterior wall with limited trouble and cost. Also you eliminate the cost of the initial major excavation.

BELOW GRADE LEVEL

The *below-grade-level* is, in my opinion, the least desirable approach (Fig. 14-6). The major reason is that exit and entrance would have to be up and down a set of steps, and a garage entrance would be at the bottom of a hill. This presents a drainage problem, unless a well-designed drain system is installed. In many cases an auxiliary sump pump would be required to pump storm drains up to a natural drain level. In almost any situation where the house is built below a level grade, the drainage system becomes a major endeavor and also very expensive. Consider the possibility of a clogged drain or power blackout. Your house would be flooded. Not a nice thought, but possible. In addition, the pump on this system will require occasional maintenance. If these features don't bother you, then go ahead and use this method.

The construction is basically the same in methods one and two. If there is one benefit to the below-grade-level house, it is that you will not have to haul in additional fill dirt when you are doing the final grading.

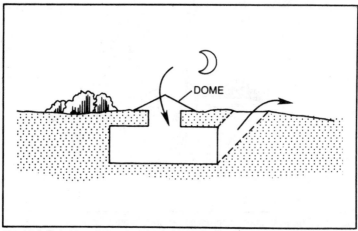

Fig. 14-6. Below-grade level method of construction for an underground home.

HALF-AND-HALF

Half-and-half is a term I've given to the many homes built, as the title indicates, literally half underground and half above ground (Fig. 14-7). At first thought, you could argue that there is no difference between these homes and thousands of conventional homes that have used their club basements as a daily living center. I guess this is technically true, but if you look closely, you will have to agree that there is a real difference. Figure 14-7 shows the major feature that qualifies the *half-and-half* house to be called an underground home. Usually the structure above grade level is used for a garage, storage, an area for solar heat storage tanks or a possibly as a work shop. In all fairness, the half-and-half homes researched for this book had less than 20 percent of the actual floor space above ground, and that 20 percent was always a non-living area. One of the advantages of this style is that exits and entrances are more conventional, thus easier to meet building codes on local restrictions. If there are disadvantages, it would be the extra cost of working on two levels and extra precaution that would have to be taken insure against water seeping along the first level into the lower level.

112

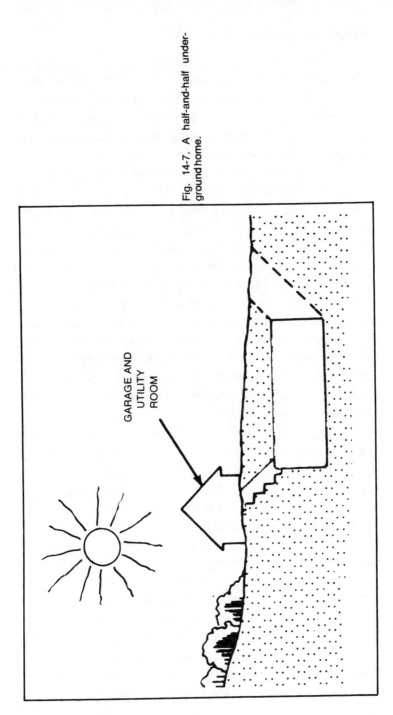

GARAGE AND UTILITY ROOM

Fig. 14-7. A half-and-half under-ground home.

BASIC NECESSITIES

Now that you're getting down to serious business with this geothermic home idea, don't overlook the forest for the trees. You're expecting those unusual problems to show up when you do something as different as this, but don't forget the basic necessities that should be considered regardless of the type of house you are building.

The accessibility to good roads in winter or rainy season should be considered seriously. It's easy to say you like remoteness and getting back to the wilderness, especially when the weather is good and you're excited about a new project and are physically active, but this house will be there a long time and so will the mortgage payments. Stop and think how you will feel about a specific location in 10 years. You've settled down a little, probably have a few more grey hairs and the children are growing like weeds. That beautiful, picturesque, four-acre tract of land you got real—cheap 10 years ago because it was 5 miles to the nearest paved road now presents the problems of daily trips to the store and children walking to meet the bus. This may not be the ideal location you first thought it was.

Also check into the cost of running telephone cables from the road to your house. The cost varies from location to location, as does the method by which the electric company charges you to bring power to your house. The way it works in Harford County, Maryland is like this: the electric company will install service lines to the meter, regardless of the distance from the existing power source at no charge to the home owner, providing the meter is installed on the nearest possible corner of the house. If you want the meter on a wall or post farther away than the nearest point, you have to pay the electric company so much per additional foot of power line required. Ask precisely what the charges will be. You will find the telephone and electric companies very cooperative organizations when planning the location of lines, meters and the place of line entry into your underground house. I won't go into gas lines because I emphatically suggest you don't even think about liquid gas, propane or natural gas as a fuel. As you know, a leak in a gas line is a serious problem anywhere, but in an underground house, it's more of a concern because of the air tightness of the structure. Don't ask for trouble with these fuels—building underground will present more than enough problems. If you stick strictly to electricity as the utility and a wood or coal burning stove for heating assistance, you'll be safer and happier.

Before signing your life away for a parcel of land, make sure water is available. Most likely you will drill a well, probably deep if you have a good piece of ground. One of the regulations in Harford County, Maryland is that an approved well usually must be drilled before construction can begin. You would be in sad shape to get your money invested in construction and find out there is no water down below. It's a necessity of life.

Well drillers are usually full of valuable information and they can probably predict the depth, quality and quantity of water you will find fairly accurate. Experience has taught them a great deal. So have faith in your well driller. You have the final say as to the location of your well, that is after the health department suggests a particular area. Make sure the well driller agrees with your choice. One important phase of drilling a well is to discuss before hand with the driller the method of payment. Most drillers have a set fee per foot for soft dirt and another for hitting rock, or they will quote you an average of the two prices, regardless of what they strike on their way to water. Also ask them about drilling a second or third hole, if the first or second turn out to be dry. Some drillers do the second drilling for half price. Some have other arrangements. Find out before you start drilling, not after you hit rock.

Once you get water in, you have to get rid of the waste it creates. So sewage disposal is just as important as finding water. Once again the local regulations dictate your septic system's location, size and configuration. Make sure your land is approved for a satisfactory system before you build, even if it is not a standard requirement. Ask for a perc test before buying a piece of ground for a private home, if public sewers aren't close enough to hook into. If the soil is not suitable for a good sewer system, it most likely is not going to give good drainage for your house.

Don't buy property simply because it's cheap. If you do, it will come back to haunt you. Sooner or later you'll pay for buying a marginally acceptable piece of ground. Remember, you don't get something for nothing. Last, but not least, check out all codes, regulations, ordinances and zoning laws that may apply to your choice of property to see if they will cause you a problem.

Some land has restrictions attached to the deed from previous owners. For example, years ago a dedicated farmer could have stipulated in his will that his farm land never be used for anything but farming. This can be done legally, and

Fig. 14-8. Be sure your land is zoned for a single family underground house.

all of a sudden you could find yourself raising a herd of cattle instead of building an underground home.

Another source of restrictions comes from development preferences. If a housing development has a set of restrictions drawn up, they may include the size of the house, material used and style. I have never heard of one that allowed underground homes, so watch out. Zoning law will be the easiest to comply with. They simply control things like multi-family dwelling, commercial ventures or farming. It's very easy to find out if a particular piece of land is zoned for a single-family underground house (Fig. 14-8). Just stop by the county zoning department and look at their zoning map. I think every county in the nation has one.

Chapter 15
Every Underground
House Needs A Skylight

The word *dome* is a term I have adapted in my writing terminology, and I admit that it is by definition technically wrong. *Skylight* would be a more accurate word. A dome is spherical in design, meaning it has all round sides; none are flat surfaces (Fig. 15-1).

Starting here and now, however, I'm going to correct the constant error of my previous book. From now on, if it is a real dome, I'll call it that, but if it's not it will be a skylight. I consider a skylight any covered structure that lets light into a house from the top side. In this chapter I will go into details of how to build different designs and how they adapt to different floor plans. Every underground house should have at least one skylight. (By definition, domes can be skylights, but skylights are not necessarily domes, if anyone really cares about the exact terminology. More exactly, every underground home should have only one skylight. The reason for this is very simple. If you are going to have a leaking problem, it most likely will be around the skylight, and it is very difficult to stop water from leaking once construction is complete. So if many separate skylights were designed into the house-roof structure, the chances of a problem increase proportionally.

Fig. 15-1. A pyramid-shaped skylight.

I can't agree with the theory that many smaller skylights at different locations (spreading natural light more uniformally) are worth the additional risk of a leak developing. Once again, this is a choice you, the builder, will have to make. Consider that it is not only the light coming in that the skylight will allow, but a window of some sort for the homeowner to look out. I can assure you that looking upward and out into my skylight eliminates any closed in feeling that you might expect by being underground. My opinion is that one good sized skylight is better than many smaller ones. Also, you must remember that every seam and/or joint must be sealed with some type of sealant. If your skylight is similar to most, you will have "Lexan" Plexiglas, aluminum, with probably a silicon or latex caulking. By the heating of the sun and the cooling of the night, these materials expand and contract at much different rates. This constant day-in, day-out expanding and contracting works any sealant to the fatigue point, and a leak develops.

LEAKS IN SKYLIGHTS

My skylight leaks occasionally, but only slightly. The leak constantly moves from joint to joint by hidden forces I have yet to identify. I figured that if it leaks at one spot tonight and I go topside tomorrow, locate it and seal it, that would then end the problem. Wrong. The leak would just move to another spot. After about a year of this hide-and-seek game with Mother Nature, I decided to ignore the leaks unless they got to be a major problem. It is two years later and the leaks haven't gotten any worse. Once I gave up trying to stop the leaks, they seemed to slow down. So Mother Nature and I are at a stalemate, and I'm content to live with a few minor leaks (in the skylight only). As a matter of fact, the leaks cause absolutely no serious problem, except that the rain droplets falling from as high as 20 feet hitting the earth below cause dirt spots to constantly cloud up my sliding glass doors (Fig. 15-2). If this type of problem would bother you, I suggest you don't consider living underground. If it's not this little problem, it will be something else to which you are not accustomed. Something will always pop up to surprise you.

HIDDEN SKYLIGHTS

One of my friends brought this idea to my attention (Fig. 15-3). We decided to call it the "hidden skylight," for the not so obvious reason that from the air you couldn't see that it even existed. The more I thought about this idea, the better I liked it, and I will suggest it to the next person that asks me

Fig. 15-2. Dirt spots on my glass doors pose no real problem.

for advice. This type of skylight has many points that make it an attractive alternative to conventional skylights. The biggest fact in its favor is that it will not leak if constructed properly. This method of construction is complex and costly, but if you really like the idea, the cost can be justified. This style must be built into the roof structure as the concrete is poured, not added on after the roof slab is set up.

This hidden skylight definitely requires a lot of extra form building preceded by a good deal of design time. I can't express the importance of this one time pouring. By important, I mean you are literally wasting your money if you don't do as I suggest. Remember, new concrete will never stick to previously poured concrete as tight as you might think. If you want to test that statement, mix up a small batch of concrete, pour it on your existing sidewalk and let it harden for a few days. It will appear that you now will have a lump in your good sidewalk forever. Wait as long as you want, one week or one year later. Now go out with a sledge hammer and chisel. With a little effort, the sample section you poured for the test will break away from the older cement, especially if it sits there through the freezing-thawing cycle of a Maryland winter. Chapter 17 has more about properties of concrete.

The hidden type of structure begins as the steel re-bar is laid and tied in place (Fig. 15-3). Steel re-bar is the heart of this structure, and it will have to be designed and located by this professional engineer that I keep telling you to work with. The most important thing you are trying to accomplish is to prevent a crack from developing at point "A" in Fig. 15-3). If the design and construction are completed properly, water will not be able to find a way to the interior of this type of skylight.

FORMS

The forms required to finish this phase of construction are no more critical than anywhere else. Of course, this

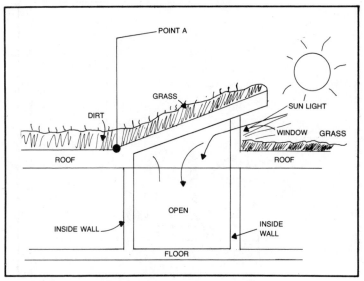

Fig. 15-3. Open one-sided skylight.

hidden style that I'm describing has unlimited possible variations (Fig. 15-4). The only limit is your imagination and your cash flow. There is one other idea I would like to get across to anyone planning an *atrium* or indoor garden covered by a skylight or dome, and that is to try to put in a setback area or, as someone called mine, a catwalk. See Fig. 15-5 to see what I mean. The greatest benefit derived from this area is that it is an extra growing place for plants that can stand extreme heat. Also, it is another area for decoration by setting flower pots, bottles or vases all around the edge (Fig. 15-6).

OPEN FOUR SIDES STYLE

The advantage of the open four-sided skylight style should be obvious. If all four sides are open, it is much easier to control air flow by opening and shutting vents depending on the direction of the wind. Plus, the extra exposed surfaces will allow more sunlight to penetrate to the rooms below. See Fig. 15-4. The major disadvantages to this type of skylight opening are few. The biggest thing to consider with

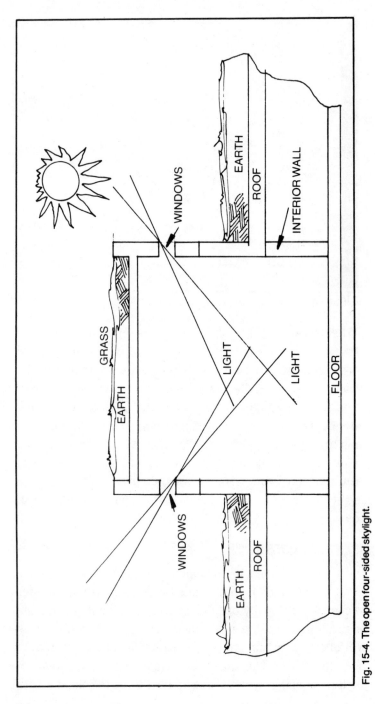

Fig. 15-4. The open four-sided skylight.

Fig. 15-5. View of the garden.

this type of structure is the fact that sunlight can not shine directly to the room below, thus affecting growing plants and heat transfer to the interior of the house if this area is being considered for passive solar heat.

However, these same points may be exactly what some underground home builders are striving for. I'll be the first to admit that at times the heat load in my atrium becomes extreme, and I wish I could cover the skylight over. Then the sun sets, and it's great to look upward and see directly into the sky (Fig. 15-7).

OPEN ONE SIDE STYLE

I just described the open four sided style, and now I'm telling about the open one side unit. This doesn't mean that open two sided and three sided skylights aren't just as feasible. It only means that the advantages and disadvantages are somewhere in between, and I'm relying on you to figure out which is best for your purpose.

125

Fig. 15-6. Plants can be placed anywhere.

126

Fig. 15-7. Typical view from the garden.

A one sided unit probably has only a couple good points worth mentioning (Fig. 15-3). One is that by situating the single opening in a certain direction, the line of visibility is cut off. This is for privacy only, if that is a problem. I'm sure somewhere somebody has the problem of a nosy neighbor. So build the closed in side facing him, and eliminate the problem.

Common sense will tell most of you that if only one side is exposed, the air flow and sunlight will be cut out proportionately. That's the way it works. You can't have your cake and eat it, too. Give all these ideas great consideration. Make sketches and reach a final decision before the house you are contemplating gets too far designed for changes to be made easily.

Chapter 16

Getting The Most Out Of An Underground Indoor Garden

Even though building inspectors, writers, contractors and other interested people call my indoor garden an "atrium," they are only partially accurate. The definition of a true "atrium" by most dictionaries is similar to an indoor underground garden, but not exactly. The definition of an "atrium" as I found it says that the top side is open to allow rainfall through. So if you care to get technical for a minute, you could say that a design similar to mine is a covered atrium. No matter what you call this area (courtyard, atrium, garden or whatever), I will proceed to explain how to get the most use and enjoyment out of this area for your underground house. In most instances, this area is limited to size. The reason is that this is the most expensive piece of real estate you own that is used for growing plants (Fig. 16-1). So you had better put this square footage to use wisely, and this can easily be done.

CUBING

Cubing is a term used in warehousing that means what the word indicates. Not only do you have the length times the width to multiply to give so many square feet of ground level

Fig. 16-1. A very expensive piece of real estate that is used for growing plants.

to work with, but you also can use the height to great advantages. This could also be called *tiering* or using tier levels (Fig. 16-2). Tiering can be done in numerous ways in an interior garden because the height in most cases is greater

Fig. 16-2. Take advantage of overhead space.

Fig. 16-3. Overhead plants.

than the 8-foot ceiling height found inside a home (Fig. 16-3). In my house, the height I have available to use is 21 feet from ground to peak. If I hang plants by using shelves and other plant-hanging devices and manage to get four different levels, I theoretically could have almost four times the original floor space available for growing plants. This extra space is a real advantage if you like gardening and a place to try your hand at year-round vegetable production. It is very possible and there are many books available on this subject. Look in the public libraries under hot houses or greenhouse growing. You'll be amazed at how easy it is to grow vegetables and flowers year-round. This idea seems to go hand in hand with underground homes, since the person interested in conserving energy usually is interested in self-sufficiency. Therefore, growing your own food is a natural (Figs. 16-4 and 16-5).

HOW TO CREATE TIERS

There are two or three easy ways to create more than one level to use for grow or show. The easiest way I know is

Fig. 16-4. You can grow many types of plants.

the method I used (Fig. 16-6). As you notice, I have created a new level over top of the existing living space. By using a ladder to climb to this level, a whole new world, as far as growing goes, is opened to you. This higher level is different for many reasons. The biggest advantage or disadvantage,

Fig. 16-5. Another view of the garden.

depending on your outlook, is the fact that the temperature rises faster on the upper level than it does below. It also becomes more intense than on the lower level. This could cause you problems if considerable attention isn't given to the planning and care of the plants at higher levels. For example, I planted tomatoes on one of the upper levels and the plants produced tomatoes in early May, but by the end of May the tomatoes were frying on the vine. Well, it didn't take me long to devise a way to shade my precious tomato plants. However, shading the plants with a sheet of polyethylene only prolongs the inevitable a few weeks, and by July 4 the heat was once again frying the tomatoes on the vine.

Here is the end result of this experiment. You can get early tomatoes or anything else you plant, but once the peak of the summer sets in they will burn up.

HANGING BASKETS

If you carefully plant vegetables or flowers in any of the commercially available pottery products and hang them from beams (Fig. 16-4), you can rearrange them as the season changes or the temperature increases. Plus they are easy to water and trim. Change varieties as you wish.

Fig. 16-6. Use a ladder to climb to the new level.

Fig. 16-7. Cactus grows very well.

IVY

One of my favorite plants for year-round growing is good old ivy. It's durable and grows fast (Fig. 16-6). It can be trimmed to suit the location. Try it; you'll like it. Ivy makes a great addition to the greenery in your underground house.

CACTUS

Cacti are a hearty species of the plant family, and they are suited to an interior garden such as mine (Fig. 16-7). They most likely will prosper in any environment similar to an underground garden. The public library and most bookstores are loaded with books on the care and idiosyncracies of cacti. I like to grow cacti, but I know very little about them.

WATERFALLS

One of the most unusual things in my garden is a waterfall. It appears that the water is coming from under the ground and flowing to the drain, but in reality I'm utilizing a circulating pump and using the same water over and over

Fig. 16-8. An underground garden lends itself to many possibilities.

Fig. 16-9. Plants are truly beautiful living things.

again. These pumps are available in any garden supply shop. This waterfall gives one more visual effect that gives the impression that you are really outdoors. This waterfall idea is unlimited in the variations that you, the builder, can dream up. All you need is a small pump. A waterfall is a natural in an underground home.

SUMMARY

An underground garden is no more than a small piece of real estate protected from the elements year-round, so it lends itself to endless possibilities. Plants are some of God's greatest creations, both in functional use and cosmetic value. Use them to the fullest extent. See Figs. 16-8 and 16-9 for additional garden scenes.

Chapter 17
Facts About Concrete

I keep writing and talking on the subject of underground homes, and I constantly tell anyone I discuss the subject with to let their engineer handle the technical decisions and design criteria for all concrete mixing, pouring, finishing, etc. Since odds are that you're not a concrete expert, this is a typical example of where a little bit of information could be dangerous. Like so many subjects, a little bit of information will give you false confidence to make decisions you wouldn't dare make if you were an expert in that field.

Concrete just happens to be one of the things that can be costly and dangerous. I have to admit that the questions about concrete, its characteristics and use constitute the majority of questions I receive. So by public demand, I will take a chapter of this book and dedicate it to detailed and specific information about the subject.

You will notice that this is probably the most specific information I will ever include. The reason is two-fold. First, I'm not the technical expert on all the subjects I sometimes lead people to believe; and second and foremost is that the public library is loaded with books on any specific subject, such as "plumbing," "electrical" and, of course, "concrete."

Even though this information is available elsewhere, I will explain as much as I can about concrete in terms the average person will accept.

COMPOSITION

Cement is a powder formed by pulverizing *limestone* and *clay* into a powder in proportions approximately 4 parts limestone to 1 part clay. Approximately 1 percent of a bag of commercial cement is other natural elements. Concrete is sold by the 95-pound bag (approximately) by almost any building supply store. The price will vary as much as 50 percent depending on place of purchase and quantity of bags purchased. So check around and save a few bucks.

Cement is often confused with concrete. It shouldn't be. Cement, mixed with water, sand, rocks, stones or any other accepted aggregate, forms concrete. You make concrete out of cement, but you can't make cement out of concrete. Now you know. When you keep hearing and seeing the words *portland cement*, that doesn't mean it was made in Portland, Oregon, or that it is the name of the manufacturer. I thought for years that the "Portland Company" bagged cement. I thought for sure it must be a big company because every bag of cement had the name "portland" on it. I was wrong. It took a little research before I found that the term "portland" comes from the Isle of Portland, England—not because the raw material was discovered there, but because the inventor thought his mixture (now called cement) looked like it came from that isle. This was around the mid-1840s. The first manufacturing of concrete in this country was around 1870 in Pennsylvania.

This is enough background in concrete composition because this is a book basically about underground homes, not concrete. But you should be knowledgeable about concrete if you live underground because your life depends on it.

AGGREGATES

The term *aggregate* refers to any rock-like particle that you add to a bag of cement to make it go farther. For example, if you have a gallon of liquid cement and a gallon of small rocks, and mix them together thoroughly, you will form approximately 2 gallons of concrete. Simple, isn't it? Aggregates can really be anything in the rock classification—any stone, slag or even pieces of pig iron, if available.

Concrete with aggregate is almost always used when pouring concrete. It's about 90 percent as strong as pure cement with no stones added, but its volume is increased two or three times depending on the amount of rock added.

Another aggregate is *sand*. Sand is always mixed with cement and rocks or stones to form concrete in proportion of about 3 parts sand to 1 part concrete.

MORTAR

Mortar is nothing more than cement mixed with fine sand and water. It is used when laying block, brick or stone. The sand and cement are mixed in ratios of about 3 parts, sand to 1 part cement. The more cement, the stronger the bond. However, a mixture with a ratio of three parts sand to one part cement is nearly as strong as an all-cement slurry, but much less expensive to use because cement is more expensive than sand.

ADDITIVES

This subject would take volumes to effectively cover, but in short I'll tell you that there are numerous additives or chemicals that can be added to concrete or cement, as it is in a slurry form, that will help or insure it to do what you want. For example, additives will allow cement to harden faster or slower, bond tighter, change colors, have higher tensile strength and more. This subject is too detailed for me to research and it serves no advantage to you. Ask a profes-

Fig. F-4. A close view of poured concrete.

sional if you are interested. I, at least, brought it to your attention.

INTERESTING FACTS

Concrete gets stronger as it gets older. I can't think of any other material that makes this claim. Concrete can also be poured underwater. This is quite a feat and requires real skill and special equipment. I mention it only because it is interesting. Concrete will also assume the texture of the surface it is placed against as it cures. A smooth finish will repeat on the surface of concrete. A smooth shining surface will likewise produce a similar shine on the concrete once it sets up. So you can control the finish of your concrete by the material out of which you build your forms.

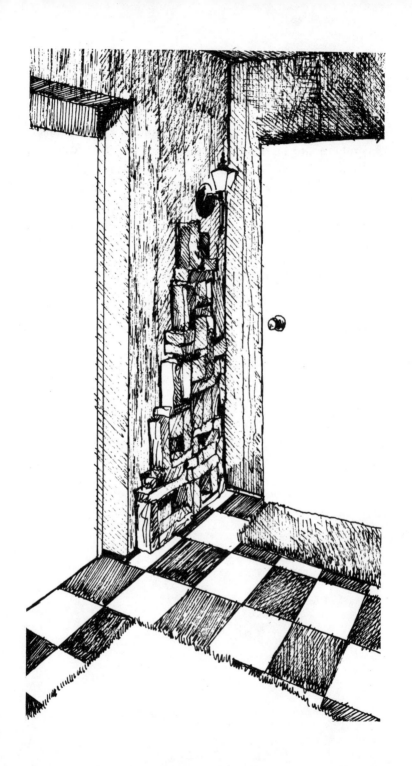

Chapter 18
Underground Flooring

How many books have you read about underground houses that have had a section devoted just to floors? Not many, I'm sure. That's because 99 out of 100 of the underground homes in the United States use the concrete slab as the primary floor. When I say primary, I mean that only carpet padding and good quality carpet are used over the concrete. But there are other approaches to flooring that will result in a satisfactory installation.

RAISED FLOOR

This method should be self-explanatory, but I'll describe it briefly. Many people have a phobia about using concrete as a floor. They say it's too hard and causes extra fatigue on your feet and legs if you stand for any great length of time. That might be true, but I think it is more psychological than physical.

As I have mentioned before, I have the best quality and thickest padding under a very good quality carpet. This combination over concrete gives a satisfactory cushion even for small children when they fall. You just can't feel the difference between my floor and that of a conventional home.

However, if you still want to have a wooden floor, here is how to do it. Once the concrete slab is poured and leveled, and all concrete or block walls and steel supports are in place, you are ready to begin. Use 2 × 4s as long as you can reasonably afford them. Many non-professional builders (do-it-yourselfers) don't realize that 2 × 4s can easily be bought in lengths other than that of the standard wall stud (approximate 8 feet long). The 2 × 4s can actually be bought in 2 foot increments up to 20 feet long. Do-it-yourselfers should also realize that 2 × 4s can be bought in lengths of 92-⅝ inches; these are called "pre-cuts." They are used primarily for interior walls. If you take a look at Fig. 18-1, you'll see why they are sold 92-⅝ inches long. They are real timesavers and cheaper.

The reason I said to buy your 2 × 4s as long as is reasonable is because they get progressively more expensive per foot. For example, a 16-foot 2 × 4 is not twice as expensive as one 8-foot 2 × 4. It is much more expensive. The reason is simple. It's difficult to handle and it's harder to

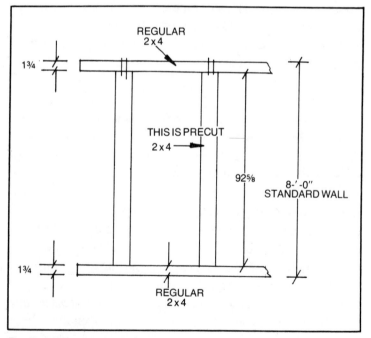

Fig. 18-1. Use precut 2 × 4s.

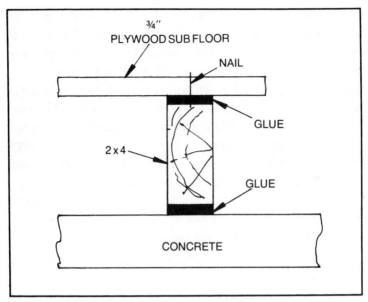

Fig. 18-2. Use glue to help the floor from squeaking.

find good timber in long lengths. Also, these 2 × 4s are not in as great a demand as the basic 2 × 4, 8 feet long.

The 2 × 4s laid on concrete covered with a sheet of subflooring will most likely squeak or creak. The longer the span between joints, the less likely a noise is to develop as you walk across this type of floor. There are a couple of things that a builder can do as construction begins to try to eliminate these noises caused by wood against concrete.

GLUING PROCEDURE

The easiest and most important thing to do would be to run a bead of construction glue along the 2 × 4s as they are laid down. This bead will act as a shock absorber and as insulation (Fig. 18-2). I strongly suggest you do this if you are laying a wooden floor (but remember, I didn't suggest a wood floor). I've used many cartridges and many different brands of construction glue at different prices, and the glue produced by the Gulf Oil Company is still my preferred brand.

Once you have the glue and 2 × 4s, begin laying a pattern of wood, never spanning more than 16″ (Fig. 18-3). If you want a better floor, move through to 12-inch centers—occasionally or about every 4 feet put a cross brace out of scrap 2 × 4s. Once the entire floor area is covered, begin laying ¾-inch thick plywood down. But first, use the same gluing procedure as before, except this time it is to stop the plywood from squeaking. Nail this plywood down very thoroughly; every 6 inches would be good. Over this ¾-inch plywood, you could now put padding and carpet, or in certain rooms it may be preferable to continue with hardwood floors or even tile. From this point on, the floor surface and material is exactly the same as in any other type of home.

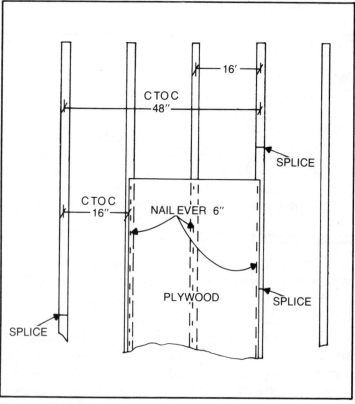

Fig. 18-3. A wood floor system.

INSULATION

The heat loss through a concrete floor with carpet and padding is around 10 percent of the total loss. This is why I say that a wooden floor won't help much as far as energy saving goes, because not much energy is lost to the floors due to the fact that heat rises. Of course, if you personally want to add insulation under the wood floor, it is as simple as "ABC." Just lay it down before the plywood is put down. Blow loose insulation in the ends later.

ADVANTAGES AND DIS-
ADVANTAGES WITH THE WOODEN FLOOR

If this raised floor has one advantage, it is not comfort. It is convenience. This area underneath (between plywood and concrete) is ideal for running small drain lines and water pipes. Electric lines can also be run under the floor, but watch the codes closely. Most locales have something to say about electric wires in this type of space.

There are really two disadvantages. The first is the additional cost of labor and material, and the second (as of yet unmentioned) is the possibility of this floor becoming a fire hazard. One of the advantages of an underground home is the absence of fire hazards.

Chapter 19
Minor Problems of
the Underground Homeowner

Every time I've ever written or talked about the design of underground or earth-sheltered homes, I've stated that I favor the interior atrium (technically correct), or interior garden, as it will commonly be called, to be included in the design. I really believe most designers will agree with me and include one similar to mine. The reason for mentioning this subject is because it is the source of a real problem. Since sliding glass or patio doors, as they are commonly called, seem a natural for doors that close adjacent to a garden, that's the way I went (Fig. 19-1).

SLIDING GLASS DOORS

These type doors violate all building codes when used as a principal means of entering or exiting a room. So right off the start, you're in trouble. Violations are not the minor problem I'm leading up to. Violations could be considered major problems. When you have sliding doors like those shown in Fig. 19-1, they are, of course, mass produced commercially. The accepted use for these doors is as an alternate exit from a room. Stop and think if you have ever seen a conventional house where a sliding glass door is the

Fig. 19-1. Sliding glass doors.

main door. If you have, it is either illegal or you're somewhere besides Maryland.

The design of the rollers and tracks for these sliding doors are for a limited number of rolls back and forth. In all fairness to most manufacturers, they are designed well for their intended occasional use. If you use the patio doors as the only means of entrance and/or exit for a room, you can imagine how many more times those little rollers rotate in their aluminum tracks. I estimated that they roll 10 times more often than their original design was intended for. The end result is that both rollers and track wear out.

Look closely at Fig. 19-1 and you'll see what my family room door track looked like after two years. This is a minor problem because the manufacturers indicate that I can replace the track and roller at a minimal cost, and I know the labor is also minimal. There is not much you can do about this little problem, except have the rollers and track specially designed out of a different material other than aluminum and nylon or plastic. I guarantee that this cost would be prohibitive. Next, you could use other more conventional swinging doors, but if you think about it, sliding glass doors are very convenient and attractive.

In addition to the actual overusing of the door for its intended use, there is another factor that plays a small part. Since my garden has sandy soil for good plant growth, it is very easy for it to get maneuvered into the aluminum track. Once there, of course, it acts like sandpaper. Everyone building a design similar to mine will have a problem of tracking dirt—greater or lesser, depending on the traffic flow. Now you know. Live with it or find your own way to eliminate it.

CURTAINS

Because this garden area gets hot, bright and dark, and for reasons of privacy, it is obvious that some type of curtain is necessary. By using these doors as a principal means of entrance, a set of hanging drapes or curtains will become soiled and dog-eared much sooner. Remember, you go in and out of these doors approximately 10 times more frequently than they were meant to be used, and those curtains and curtain rods were never intended to move. See Fig. 19-1 for another reason why curtains get worn out fast.

Chapter 20
Major Problems of
the Underground Homeowner

I have actually lived in my underground home for 2½ years as of the writing of this chapter. A better chapter title might be describing problems that could "aggravate" you.

CRACKS IN THE EXTERIOR WALL

I agree that the words "crack in the wall" might be grounds for alarm when that wall is all that is separating cold moldy earth from the inside of your living room. You then can imagine my thoughts when one of my children very nonchalantly said, "Dad, did you know the wall in the dining room has a crack in it?" I immediately pulled my set of blueprints from their highly secure storage file (under the bed) to locate what could possibly be going wrong. I thought, "When will that wall collapse and the entire 1,000 tons of debris bury me forever?" My memory was really accurate because the prints confirmed that at the point of the crack on the interior Sheetrock wall, the concrete wall behind it was designed well and, in fact, if that wall had a weak point, it wasn't at this location. Nevertheless, a crack about 3 feet long had developed. The only way to find out for sure what was causing it would be to remove the Sheetrock in that area.

What I found was a hairline settling crack between the concrete block. There is no explanation for why it settled at that point rather than somewhere else, but it did. Don't get alarmed. It causes absolutely no problem that I can tell you about yet, and I don't think it will.

As a matter of fact, the crack in the Sheetrock interior wall was greater than the one in the block wall. I watched this crack closely for about six weeks, long enough for a few heavy rains to fall. I proved to myself that it did not leak (I didn't expect it to). Once I was totally convinced, I reinstalled the Sheetrock, finished sanding the wall and repainted it. The crack that once caused me concern has been eliminated and shows no signs of returning. It was nothing more than an initial settling crack that will occur in any house, conventional or otherwise.

The point you must not take lightly is that in an above ground structure this crack could almost never be a major problem, unless the builder intentionally screwed up. But in an underground house, even the best design and construction methods could cause a little problem that would be difficult and expensive to correct. So if you are one of these people who go through life with a black cloud always over your head, you had better live in a standard condominium. It's safer.

CRACKS IN THE ROOF

A crack in the roof slab or as you see it, a crack in the ceiling is a cause for real concern. If you built your home as I recommended in my first book, the ceiling in your home is really the underside of a slab of concrete. I still recommend that you do not install a sub-ceiling over the concrete slab for this one reason. You can always observe what is going on as far as potential problems go. I also discovered a hairline crack in the slab overhead. I thought I had a problem, so I consulted with my personal concrete expert and he assured me that it was only a curing crack, and only on the surface. The crack did not penetrate through the slab. These type of

Fig. 20-1. A roof is strong enough to hold cars.

cracks are expected and should not cause concern. I painted over them and that was the last I saw of them. Now if that crack continued to reappear and got bigger, then real trouble is ahead. I will not go into details about what to do because if this is one of the problems you face, you're going to need more detailed advice than I could put in this book, plus you need an in-person expert, not a mail correspondent expert. If the roof slab is laid down correctly, you never have to worry.

ROOF INVADERS

Did you ever stop to think that in most underground homes anyone could drive anything up on your roof? It sounds strange, but it's true. I realize you could put up a fence, but excluding that it's true. One of the questions teenagers constantly ask is, "What if I drove my car on your roof?" So I have to prove my engineering and construction ability by taking my car across the lawn and showing them that it is strong enough to hold a car (Fig. 20-1). I don't want to prove the point over and over again because I'd rather not push my luck but I have had as many as four cars on the roof at one time. Once a visitor sees that it is no problem, the discussion never comes up again.

A problem that you do have to be aware of is that an underground house always has some place for people or animals to fall from. See Figs. 20-2 and 20-3. These ledges or walls can be protected by a fence or barrier of many types, but they are easily avoided if anyone really wants to climb around on one (Fig. 20-4). This is a condition that any

Fig. 20-2. Rooftop view.

Fig. 20-3. Another rooftop view.

underground home will have and you have to accept it. Children are the greatest offenders due to their natural curiosity (Fig. 20-5). It is literally impossible for a small child to get onto the roof of a conventional home without a

Fig. 20-4. A place of potential danger.

Fig. 20-5. View of a pyramid skylight.

ladder. With my house and others like it, there is nothing to it. My insurance is paid up and I watch children closely when I know they are around, but children, being the curious, adventurous humans they are, will try to climb anywhere they shouldn't be. And I don't mean just climbing. My son is 13 years old and he and his friends have been known to ride their bicycles over the side. I realize this condition isn't always going to be there if your design is different from mine, but odds are that there will be a similar condition (Fig. 20-6). The situation in Fig. 20-6 is dangerous, and the photo was staged to prove the point. I think I've convinced the neighborhood daredevils that my roof is not a bicycle obstacle course.

When winter set in and it snowed, the first thing I discovered was that my rooftop had been designated as the neighborhood "downhill Junior Winter Olympic sledding and skiing course." It didn't take long to realize that I was letting myself in for real trouble. If someone got hurt, and a sled is dangerous weapon on hillsides like the side of my house, I would be responsible. Needless to say, I put a halt to the

Fig. 20-6. A dangerous bicycle ride.

sledding. The remainder of my property is great for sledding, so I compromised with everyone and permitted use of the driveway for sledding, since it also is a big hill, only safer.

I did agree to let children play on the roof and the side banks by using it as a slide. I consider this the safest alternative, and it can be practical in the summer and winter (Fig. 20-7). If you question what they are using to slide easily on the grass, it's only a piece of cardboard. It's free, relatively safe, and they get great exercise for hours at a

Fig. 20-7. Kids like to slide down the bank on cardboard.

time. The only harm that can be caused is they have been known to dig their feet into the ground to stop, and this tears the grass away. If I see this wear and tear, I ask them to use another spot, and they always comply. I think they appreciate the unusual use of my home. Living underground has some strange advantages or problems depending on how you look at it.

CHIMNEY LEAKS

Chimney leaks have never caused me a problem that would be considered major, but they have been a constant source of aggravation. It has been an expensive series of trial and error methods to solve the problem. My first design I admit was doomed to failure as I take a closer look. Hindsight engineering, like bar room quarterbacking, is a tremendous advantage. You really see things in a different light. My intial approach was basic but bad (Fig. 20-8). My second approach wasn't much better (Fig. 20-9). By the time I discovered that a third method of construction was going to be needed, I was

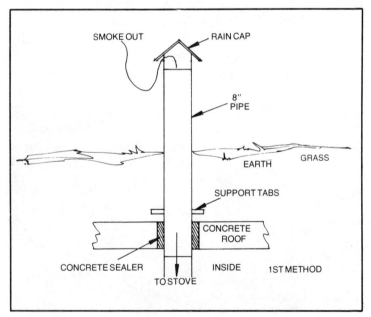

Fig. 20-8. Do not use this method to design a chimney that won't leak.

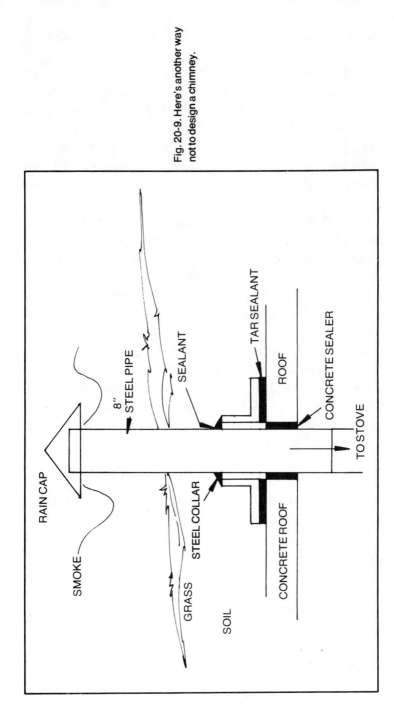

Fig. 20-9. Here's another way not to design a chimney.

RAIN CAP

SMOKE

8" STEEL PIPE

SEALANT

TAR SEALANT

ROOF

CONCRETE SEALER

TO STOVE

STEEL COLLAR

GRASS

SOIL

CONCRETE ROOF

irritated with myself. (See Fig. 20-10 for the best method). I couldn't believe that I could do the monumental task of building my house, working at a full time job, holding my own against the regulation agency (building inspectors), writing my first book and raising a family while still not designing a chimney that wouldn't leak. It is somewhat like admitting that you won all the battles but ended up losing the war. Needless to say, my third attempt at designing the chimney was given a lot of thought.

You must realize by now that this is the fifth time I've shoveled the dirt in this area, counting the ground breaking, the first grading and two previous attempts to stop the leaks. I feel I know some of this dirt personally by now. So you might be saying persistence finally paid off. You would be wrong. My third design, even though 99 percent effective, still let droplets of water drip down onto the hot stove and immediately evaporate with a sizzle. It definitely was not a problem that could not be tolerated; it was just an annoyance. Finally I developed a ring of aluminum foil that catches the droplets and leads them to a catch pan. This method is still in effect today. I can not bear the thought of shoveling that dirt again in the near future. My plan is to let it go for another year, sell a few books and try again when I've mellowed a little more on the subject.

The reason for the problem is because I underestimated the power of water turned to steam by the hot chimney, the expansion and contraction of the steel chimney, and other actions of nature and laws of physics. The heat causing expansion is the biggest barrier breaker. I know this for a fact, because I do not have one drop of water at the chimney if the wood stove is not operating, even in the spring with heavy Maryland rains. Only during the colder months does the chimney leak. I have designs that I will give a lot of thought to because I guarantee that the next time I dig up that area, it will be the last. If you are designing your underground house, recall this little problem I had and avoid it.

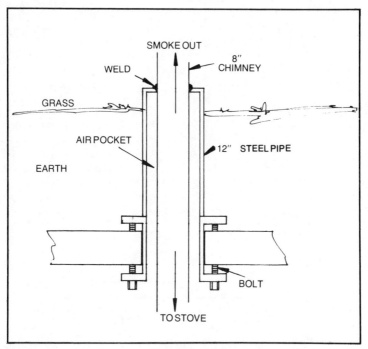

Fig. 20-10. My third attempt at designing the chimney was much more effective.

GRADING

Getting grass to take on any lawn in the springtime in most sections of the country is somewhat of a problem because of heavy rains at times, but getting grass seed to hold on a 45 degree slope when 4 inches of rain falls in one day is nearly impossible, or so I thought. In each of the past two planting seasons, part of the banks slid down, similar to a small mudslide. Each summer I would pile the dirt back up, and by fall the grass was growing again. Each spring the dirt slid back down. I was about to agree that the degree of the slope would have to be made less, which would require a lot of new work. Plus, the retaining walls would have to be changed to match the new slope.

One of my neighbors said he solved a similar problem by laying lengths of concrete wire reinforcement mesh down and staking them into the ground. Then he laid a couple of

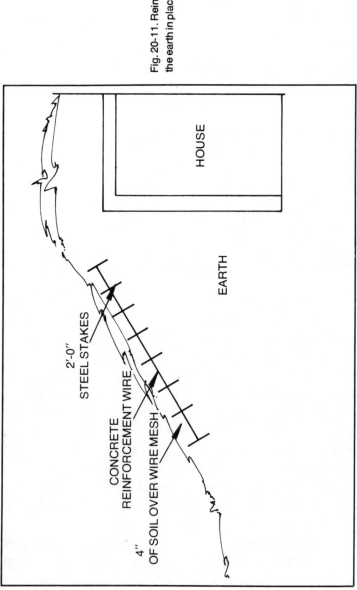

Fig. 20-11. Reinforcing wire holds the earth in place.

HOUSE

EARTH

2'-0"
STEEL STAKES

CONCRETE
REINFORCEMENT WIRE

4"
OF SOIL OVER WIRE MESH

inches of topsoil over the mesh. I tried this method and it worked like a charm (Fig. 20-11). I found that if you have a slope of 40 degrees or less, all you need is good soil, good seed and light rain (Fig. 20-12).

If your slope is 45 degrees or greater, the wire mesh is a must. Don't think that planting sod is the answer. The sod will become soaked with the first rain. Since water weighs over 62 pounds a cubic foot, it adds great weight to the already heavy sod, and down the slope it slides. Stakes only hold the sod in place immediately adjacent to the stake itself.

I would suggest that you ask a local landscaper his opinion on what you should do. Weather conditions in the growing months vary from locale to locale enough that my advice, based on Maryland weather, might not be good advice for your area. But don't forget the wire mesh; it will rust away in a few years (maybe five or so). By then, the grass will have a good root system, and the soil will be as compacted as it's going to get.

LAWN CUTTING

If your house is similar to mine and the side slopes are as you see in Fig. 20-12 there is really only one practical way

Fig. 20-12. Another front view of my home.

to cut the grass on this bank. Use a lightweight rotary lawn mower with a rope tied to the back (see Chapter 9). This may seem awkward to you the first time, but it is very easy, and efficient, as well as great exercise. I can't make this following point strong enough. Do not walk up and down or across the slope with the lawn mower. Stand at the top edge only and let the mower down slowly. Pull it back up the same way. You will soon learn to control the mower easily and also learn to avoid cutting the rope. As for the rope, use ¼-inch or thicker. Tie knots every so often to help with the grip required to pull the mower up. Of course, use gloves to avoid rope burns. For safety's sake, do not let anyone get below to watch what you are doing. This rope has broken a couple of times on me and the mower goes until it hits something, cutting most things in its path. If you think I'm over-exaggerating the safety aspect of this lawn mower, don't take it lightly. In the past 20 years, I have had two different accidents, one with each foot slipping under a rotary lawn mower on a hill—not at this underground house but at previous locations. I think I've learned my lesson. You learn by my mistake; it's less painful.

Now that you've thought about the problems of getting grass to grow and then cutting it, you might say, "If it is that much trouble,then I won't plant grass on my banks." All I can say on that subject is what I have heard landscapers say, and my logical deduction tells me that it is probably true. If you plant an ivy or greenground cover, it looks okay for a while. As it grows and the roots take hold, the leaves spread out like an umbrella. This kills anything like grass that would try to fill in the spaces between plants. When a heavy rain falls, the water runs under the leaves, washing dirt away. Once erosion starts, it is hard to stop. As I said, I have no personal experience on this problem, but I think it should be given thought. The only way you will know is to try it, because with each situation so different (slope, climate, etc.) you never are really sure what will work.

BUILDING CODES

Building codes are such major items that I have set aside the next chapter to cover my thoughts and suggestions on them. I only mentioned them here so you would realize that they are definitely a problem related to building underground.

Chapter 21
Codes and Regulations

If the problems of building underground were divided into two categories, they could easily be titled *physical* and *mental*. The physical category would, by description, include everything requiring physical labor, such as digging, hauling, building and so on. It would also cover the subjects of waterproofing, electrical, plumbing, etc.

The mental category would cover such things as worry, fatigue, hassles, arguments with opponents, financing and other non-physical subjects. The biggest mental aggravation that was not mentioned in the previous sentence is building inspectors who, in reality, are the only link to the real culprit of underground homebuilding aggravation. That aggravation is building codes.

THE REASON FOR BUILDING CODES

The best argument for some sort of building code program is not the most obvious. The most obvious, as I hear it, is that codes are to protect your physical health. They prevent you from building a fire trap or a roof that will fall in on you, or they prevent health hazards, such as sewer systems that are inadequate. All this is fine, but the reason is not a good one. The government, be it local or federal, has no right to protect you against yourself, which is what govern- ments are trying to do. So the line of, "We are only trying to protect you," is out of order and none of anyone's business.

The real reason that should be given for building codes is just as basic. Codes should be enforced to protect the "future buyer." I will agree with that statement, but then only to a reasonable point. I wonder whatever happened to the old phrase, "Let the buyer beware." There are many laws on the books to cover fraud so the buyer is really protected against an unsafe structure, if indeed he were led to believe it was a safe structure. Building codes, by original intent, probably were good ideas. However, due to the nature of government agencies, they begin to build dynas- ties. They grow bigger and bigger and everyone has to justify their mere existence at their job.

Can you for one minute imagine an inspector being called to a site to inspect a house, looking it over, and saying that it is perfect? No way. It has never happened and never will. The inspector always finds something to point out, so he will feel that he did his job. Now comes the kicker—just think of the possibilities that he has to choose from if he really wants to nit-pick. They are limitless. I only hope that any inspectors or officials reading this book will take my point in good faith and realize that they do put undue burden on the builder, many times unnecessarily.

If you intend to build in a community that has a complete set of building codes and zoning regulations, I'll remind you of a few things that will give you and them concern.

MEANS OF EGRESS

Probably the most difficult to comply with is a term called *means of egress*. This meaning exits. This covers doors, windows, holes or anything else that could be used as fire escapes. Usually the codes require two exits for each room except the bathroom. Obviously when you build underground the windows are the first to go, thus a problem.

Sprinkler System

The solution that was proposed to me was to use a commercial sprinkler system in place of the missing windows. If you price their systems, you'll soon find it unreasonable from a cost viewpoint. In my house an acceptable sprinkler system would have cost approximately $10,000. Needless to say, I couldn't accept that.

Smoke Detectors

Therefore, after much negotiations, conflict and letter writing, we (the county and I) agreed that a smoke detector in each room would be adequate. However, even on our compromises we ran into a snag. My idea of a good smoke detector was a battery-operated model made by any of the major smoke detector manufacturers. When I say battery-operated, I mean the type that doesn't require hook-up to the 120-volt house electricity. They only require small voltage batteries, usually one 9-volt battery that lasts for over a year. My logic for this type is understandable, I think. If the house electricity is out of order by storm or if a fire starts in the electrical circuit, it usually blows the fuse or breaker, thus shutting down the power to the detector that is supposed to be on guard. But a battery-operated model is on guard at all times. These battery models usually have a warning system that would sound when the battery is wearing down and ready for replacement. The county inspectors, as usual, had a different view of the same subject. They say that human nature, more often than not, will forget to replace the run-down battery so that the detector system will most likely be inoperative after the first battery wears down. I felt I was right, but I admit that it's all in how you analyze the problem.

Series of Corridors

There is another alternative to meeting egress (fire escape routes) regulations. That is by a series of corridors leading from each room (Fig. 21-1). The reason a corridor or hallway is required is because most local codes say you must

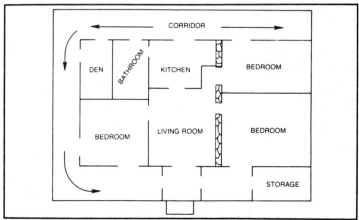
Fig. 21-1. Typical example of using corridors as a means of egress.

be able to escape to the exterior by means of a window or door, without going through another room. For example, you cannot escape out of your bedroom into the living room to get to the nearest door or window. Your bedroom must have a window that leads to the outside and a doorway that leads to a corridor, since corridors are not considered rooms by definition. Now you can easily see why underground homes get into conflict when you eliminate all of the exterior windows. To meet all codes by the corridor system, you would have to waste a lot of square feet of expensive floor space, plus isolate some living areas. Corridors are not my personal idea of a solution, though you may work out an acceptable arrangement.

In summary of this fire escape problem, you have basically four choices to consider before you finalize your plans:

- Provide a corridor to each room
- Provide an approved sprinkler system
- Provide approved smoke detectors in each room
- Provide escape from each room through the roof

It is my opinion that the third solution is a very economical and logical approach, since that is the way I solved the problem.

The fourth solution is listed more in jest than for serious consideration. I hope it's obvious that the more holes you have in your roof, the more problems you will undoubtedly have. Also, the cost could be prohibitive.

RETAINING WALLS

Another area that could cause you some trouble when trying to meet codes is retaining walls. Most underground homes have more than their share of retaining walls, depending on the design, of course. Most localities have regulations as to maximum height and method of construction. Try to arrange your design to eliminate exterior retaining walls, if at all possible. If you can't, figure the cost closely. They are extremely expensive to build if they meet the code.

If you expect a retaining wall to stand on its own when the force of mother nature, by means of freezing and thawing the soil, combined with the weight of the soil, is trying to push it over, then you are naive (Fig. 21-2). You can reinforce the wall with enough steel and enough concrete to

Fig. 21-2. A large retaining wall is unsupported in the final stage of construction.

insure that it will stand and not crack against any of the elements, but it is difficult. One of my solutions to retaining walls is to always brace one against another with steel, reinforced concrete (Fig. 21-3). If you have only one retaining

Fig. 21-3. Two retaining walls held apart by the roof structure.

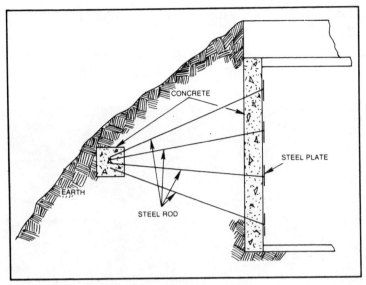

Fig. 21-4. Alternate method of holding a retaining wall.

wall, you can always tie it to an anchor post of some type (Fig. 21-4).

Additional suggestions on how to tie a retaining wall back could be lengthy and it is a subject that has been written about in detail. Ask a good contractor or engineer, or search the library for books on the subject. It will be worth your time to particularly look into the use of railroad ties. Most railroad ties which you can buy now have never been used on a railroad and they aren't really the same thing. The old-time railroad ties were pressure cooked in creosote (a preservative) to prevent rotting. The new so-called ties have only been dipped in a preservative and will not last as long as a real tie. Building retaining walls of ties is a difficult, back-breaking job. So use something that will stay in place as long as possible.

HAND RAILINGS

Another subject that will require close attention to complete your house is hand railings. There is usually a difference between safe conditions and meeting the codes. I ran into a problem with my interior garden (Fig. 21-5 and 21-6). When a garden is covered by a dome, it is a personal judgment as to whether it is considered exterior or interior. If indeed the inspector feels it is exterior, then hand rail,

Fig. 21-5. Partial view of a garden in the final stage of construction.

Fig. 21-6. An exterior garden and walkway.

doors and electrical boxes must meet one segment of the code. If he decides it is interior, then he will turn to another page in his book. As I mentioned, hand railing requirements are different depending on the use of your garden or atrium. If you plan to use an above-floor level to grow year-round plants, then you are required to have a railing, even if you only use it once a year. If it is just for decoration, then no railing is required. Still another problem area could be glass doors. The codes sometimes say sliding glass doors are not allowed as a principal means of egress. This becomes particularly complicated when these doors enter into your garden. Is it exterior space or interior space?

Most local building codes were patterned after a basic set of national regulations and sometime in the past these codes have been adopted as the gospel truth. In addition, many subdivisions have drawn up a few amendments to the national codes. Most of these amendments are pets of some local developer and in no way are intended to serve your benefit, only your expense. For this reason, everywhere in this country will probably take a different approach to the building of an underground house.

AIR CIRCULATION

Air circulation is one area that is touchy and important. If your house is like mine when complete, you will have a wood stove for heat. I'm not suggesting that you don't put conventional heating in if you feel comfortable doing it, but the majority of underground home owners find wood heating is the most satisfying, even over solar heat. If you use conventional heat with a duct system, then all air circulation problems will be easily met. But if a wood stove is in your future, then concentrate on the area of circulation. Just because you heat one room easily and quickly doesn't mean the hot air will move from room to room. You need a circulation system, but as you know a wood stove needs the same oxygen that you need to breathe. As the fire burns and you breathe, this oxygen must be replenished. So you do need a duct to the outside world. Here is where the codes will conflict with your good judgment about the size and location of the fresh air intake supply. The inspector's logic here is that in a conventional home fresh air is drawn in around windows, under doors, etc., regardless of the new caulking and installed insulation strips. I know from experience that all you need to do is circulate the air inside the house. The exterior doors would have to be opened approximately once a day to provide all the fresh air necessary to live comfortably. This is particularly true if you have a large dome or breathing skylight. Remember, if you keep bringing in outside air in the winter when it's not required, you defeat the purpose of underground living. Of course, the same is true in the summer. I like and need fresh air as much as anyone, but enough is enough. Don't change air more frequently than necessary.

HEATING SYSTEM

Do you know that a new home in most parts of the country must have an approved heating system? A wood stove won't suffice; it can only be a backup unit. If you plan on heating by wood as I do, you might have to install a nice furnace and let it sit there gathering dust. This is sad but true.

ELECTRICAL CODES AND PLUMBING

Plumbing and electrical codes will be mentioned only in passing. There doesn't seem to be much of a problem with building underground and meeting plumbing and electrical codes. Electric wires and water pipes don't know the difference so at least these two phases of utility installation should be simple (Fig. 21-7).

I will tell you to watch out for the sewer drain vent pipes. Don't put them through the roof. Run them with the water pipes along the block wall and vent them to the exterior vertical walls. These drain air vents are required to be 2 inches in diameter to allow proper air intake into the septic system. The reason these drains are required is that if you release a large amount of water into a drain line at a certain point, it creates a vacuum as it drains to a drain field. This vacuum draws the standing water out of the traps in the sinks at another point. Once the water has reached the septic tank, the trap at the second point will be empty and allow sewer odor to seep back up the pipe into the living area.

Fig. 21-7. Typical water pipe attached to 1 × 3 furring strip on block wall.

STAIRS

Still another area you might watch out for and design around is the steps. The reason to avoid steps in an underground house is, once again, written in the sacred scrolls of building codes because a principal means of egress cannot contain steps upward, only downward. Therefore, since you are already down, you can't continue in that direction. You must eventually come up. This is a code that varies from locale to locale, so check your particular situation out carefully.

DOOR OPENINGS

Most likely all doors opening to all living areas from utility rooms, storage rooms, garages, laundry rooms or furnace rooms must be made of solid wood approximately 1½ inches thick with no windows. They also must have an approved burn rate. The reasoning is that it will contain a fire in these work areas long enough for you to escape past (not through) the door.

SKYLIGHTS

You would think that a skylight would be considered an option to a house roof especially if the house is underground. You are wrong.

Once you decide to put a skylight on your underground roof, it must comply with more codes than you can shake a stick at. First a skylight must be anchored to the roof slab even if it weighs 4000 pounds. Even if it is too heavy to go anywhere, a skylight has to be anchored. I don't know why, and neither does anyone else. But the code says "anchor." So you "anchor."

Next, the panes must be of a material that meets a specific strength test, but no one that I am aware of knows how to test this material. Of course, Plexiglas or ¼-inch acrylic passes the test. I suggest you use ¼-inch acrylic for your skylight and avoid problems.

USED MATERIAL

Now you would think that any building material would be okay to build with. You are wrong one more time. Some areas have codes requiring that you only build with new material. That means no antique wood or no used brick or block, as I used. Even used 2 × 4s do not comply to certain codes. By now I think you get the picture. Building codes are here to stay, so you had better just find a way around them or comply with them. A good lawyer can always take short cuts if push comes to shove.

Try to meet code requirements, but if necessary fight them. Deviations are possible to any code. By now I think you understand that building underground isn't a piece of cake.

Chapter 22
Bermed Style

The term *berm* means a mound. Therefore, a bermed house is a house in a mound of dirt. This is a design anyone interested in an underground building should investigate first. It is by far the safest way to build underground, even though it is only technically underground. I personally feel that for a house to fall in the "underground" category, the building should be below natural grade level by approximately 5 feet.

In spite of what I just said about a bermed house, I do think they have their place and definitely have an advantage over conventional homes. They have at least one advantage over standard underground homes. That advantage should be fairly obvious to you by now. (From this point on, I'm going to call this type of house an above-grade underground house.) I picked the term "bermed" up from a newspaper article. The word by dictionary definition doesn't fit.

WATERPROOFING

Waterproofing is required on this style of structure just as on any other underground house. The thing to remember is that the pressure and volume of the water trying to find a path into your house are changed and reduced greatly from a house 5 feet underground (Fig. 22-1).

CONSTRUCTION

Tied directly to all other advantages is construction. There is no doubt that above ground construction is easier than below grade. This is due to many reasons. First and foremost is the fact that trucks can deliver building materials closer to the site as they are needed on the grade level, as opposed to the problem of physically manipulating supplies down in an excavation. Many times using a crane would be the only way possible. This could be quite a cost saver, depending on the design and the terrain you choose to build on. The deeper in the ground you build, the more expensive it will be.

Drains are a phase of construction that is greatly eased by being on grade level. Laying pipelines and spreading gravel are easy jobs if you can maneuver freely with small loading equipment and manpower, but trying to spread tons of gravel and lay pipes down in a trench as much as 20 feet below the grade is difficult at best. It is also dangerous because walls and trenches will collapse once soaked with water from a rainstorm. Above grade construction definitely eliminates this potential problem.

SAFETY

I know from experience how much equipment like shovels, hammers, nails, etc., fell accidentally during my construction. This was because my helpers were handing those tools down into the trench from 15 feet above, and accidents will happen. By building on grade, this unusual condition is nearly eliminated, especially in early phases of construction.

COST FACTOR

This is another phase to consider. Building an above grade underground house eliminates a lot of excavation. If

Fig. 22-1. Note my name and house number.

you haven't considered excavation as a big cost factor, you'd better sharpen your pencil. The increased cost of fuel alone accounts for the high hourly rates that a bulldozer owner gets for his machine. The bigger the dozer, the higher the hourly rate.

The cost of excavation and grading for an underground house is approximately 10 percent of the total cost. Of course, this is assuming your construction is something like mine. If you think this is high, let me assure you that I kept accurate records of my daily activity. It wasn't hard to go back and add up the hours and plug a dollar value to each hour. Like I've said before, I know what I'm writing about because I did the work myself and I hustled. If a contractor were being paid by the hour, he wouldn't work as consistently as I did. So my estimates of 10 percent of the total cost would be low if you were contracting the grading to an outside firm.

AESTHETIC VALUE

In spite of the advantages of building a bermed (above grade) home, it is very difficult to landscape a mound of dirt to be pleasing to the eye. It has been done, I am sure, but try visualizing hiding a mound of dirt 15 feet high, 50 feet long and 40 feet wide. It isn't easy. It is like trying to hide a wart on the end of your nose with a powder puff.

If a house is built above grade, the slope of the earth is probably going to be steep (Fig. 22-2). This makes it difficult to get a grass covering to grow and even more difficult to keep it cut or trimmed (Fig. 22-3).

SUMMARY

Give above grade homes a look and make a few inquiries, but if you want my opinion drop this idea. Underground homes are a novelty and they will always be. From the discussion I've had with potential buyers, I'm

Fig. 22-2. Note the slope of the earth.

Fig. 22-3. It can be tough to keep the grass trimmed.

convinced they wouldn't be interested if my house was a mound above ground.

Resale value is a point to consider in great depth. It could be your only retirement plan, just as mine will be.

Chapter 23
Other Alternate
Designs That Will Work

You should realize that God didn't call up one day and appoint me chief curator of all underground home operations. This is why I try to constantly remind you that most ideas included in this book are personal opinion. It is true that I have checked out any specific details I may write about to the best of my professional engineering ability. That still doesn't make the information absolutely, totally and always accurate. This holds true for any writer on any subject whether they admit it or not. It is also true that the old phrase: "Guilty by association" fits anyone in my position. It would be very difficult to design, engineer, build and live in any type of structure without becoming somewhat of an expert on that particular style. Please don't confuse my opinions on the subject of underground homes with actual proven, documented information. I'll usually indicate whether it's proven or opinionated information.

HOUSE IN A SHELL

This term *shell house* was presented to me by a potential underground home builder who wanted me to do some engineering for him and act as his consultant. The only problem with this plan of his is that he wanted me to do it for free. He said it would be good experience. I promptly told him if the concrete company would mix and deliver his concrete free just for the experience, I would be happy to contribute to his cause. Needless to say, I never heard from him again. He did however, get me to think about the "shell" idea for underground housing. I will not divulge any of his information because it wasn't very accurate, but I will expound on the possibility of this style of house. A shell underground house works on the same principle as a thermo-pane glass window. What you essentially have is an exterior wall holding back all the elements. That could be a problem. Then inside this shell you build a very conventional wooden house (Fig. 23-1). This is a good idea, although I have never seen it built.

Positive Side

This type of structure only has one totally positive aspect over most other styles of underground homes. You have the interior separated by a space. In this space you can do many things such as circulate heat, cool and dehumidify air. You can run electical wire around through it; even water pipes could follow the wall around. This space could be used for storage, but the real value of this aisle is that any exterior wall problem such as cracks or leaks will be separated from the interior, and in this space you can work to eliminate any problem. You could let water leak through the exterior wall and drain it around the interior. This is not suggested as a means of design, but if worst came to worst you would at least have a choice. In my house and most others, if the exterior wall had a major leak, you would soon know it inside the house, with no easy method of fixing that leak. It definitely is of value to visually inspect exterior walls for defects and know if they show up that they can be corrected.

Air Circulation

This space that surrounds the interior living space could be the *plenum* for all your air treatment— even your source

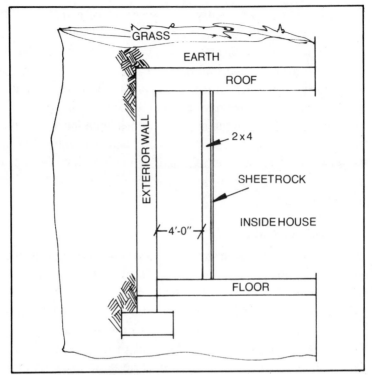

Fig. 23-1. House inside a shell.

of fresh air. This is a subject that will get very complicated and is very special to each and every home built. With the different temperature, rainfall, humidity, wind, velocity and even cleanliness of the air, you can see why every geographical location would require different specifications. This is where an engineer will earn his money and save yours. When figuring the treatment of air regardless of what type house it is, let an expert do it, or at least listen to his advice. This space could act as a duct work system allowing treated air to travel to any room without the standard sheet metal duct system usually required (Fig. 23-1).

Negative Side

One negative point when using the hallway method is the loss of floor space, or better explained as floor space you have but can't efficiently use. See Fig. 23-2 and see how two floor plans with the same dimensions of 40 feet by 50 feet, or 2,000 square feet of floor space, work out if you measure only living area. If you multiply the figures out, you find that nearly 30 percent of the space inside the exterior walls is unusable for every day activity. Since the price of homes is closely related to the square feet available for living space, you might want to reconsider this "house in a shell" design. The construction methods and the engineering are basically the same as any other style from a cost and difficulty standpoint.

FIVE-SIDED SHELL

This five-sided shell variation is not as simple as it may at first sound. The original concept was an exterior shell around the four vertical interior walls, and that idea is not really that difficult to conceive and/or build. The fifth side is the real kicker, since it refers to the roof structure (Fig. 23-3). This is also a good idea, and it will work. It is usually a matter of economics, and this style is no exception to the rule.

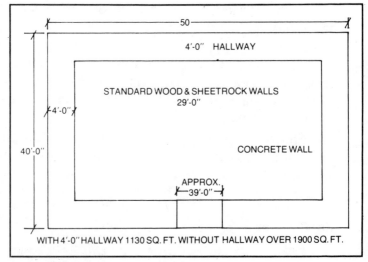

Fig. 23-2. The hallway takes up valuable space.

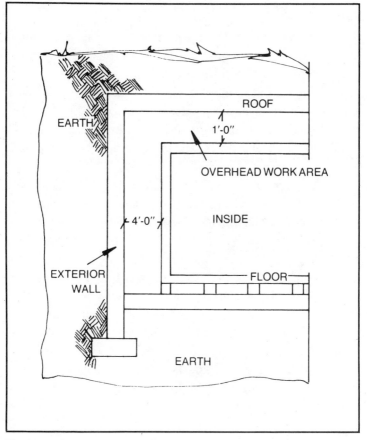

Fig. 23-3. Note the overhead work area.

Design Principle

The principle of this design is to create an air chamber over the conventional ceiling by a series of wide beams. These beams could be either steel eye beams embedded in the concrete as it is poured, or the conventional concrete beam with reinforced bar placed for strength. The overhead space will come in handy for running water pipes and electrical wiring. The air in the overhead space would have to be constantly circulated to prevent a musty odor from developing in the summer months, but that in itself is not a major consideration.

Disadvantage

The major drawback to this approach would be the initial cost. Just the labor alone to build forms and lay steel reinforcement rod would be approximately 50 percent higher than the labor for a single thickness concrete slab as most underground homes have. Also, the engineering time required to design a structure such as in Figs. 23-2 and 23-3 would be greatly increased, and this converts to more dollars very fast.

A third problem would be the use of extra equipment, particularly a concrete vibrator, to settle the fresh concrete in all the cracks and crevices. The cost to use this vibrator is not where the cost stops, however. These vibrators do a fantastic job of shaking concrete down. They do an equally good job of shaking concrete forms apart, and then a major construction disaster occurs. This is exactly why I say that the cost of form building would be semi-prohibitive to the average do-it-yourselfer. Single span roofs are always more expensive than roofs with interior bearing walls. Think about this method. Talk to the experts about it, and then draw your own conclusion as to your ability to complete this phase of construction.

TOTAL ENVELOPE SYSTEM

A total envelope system is almost self-explanatory, and I don't want to repeat and retrace the preceding information. The conditions are exactly the same as in a five-sided shell. The only difference is that the floor is raised approximately 4 inches to allow for air circulation under the floor as well as around the four walls and over the ceiling (Fig. 23-1). There is a slight benefit. By insulating the floor with this layer of air, you are saving energy. Remember, 10 percent of the heat loss in an underground house is through the concrete slab used as a floor, which in most cases is insulated only by carpet and padding. You must realize that only a fraction of this 10 percent loss can be prevented no matter what you insulate the floor with. So it is my opinion that the extra cost of material and the additional labor will never be offset by the proposed theoretical savings due to less heat lost through the floor.

If you really like the raised floor idea for its cosmetic value, due to the fact that you could now have hardwood floors, then use a raised floor by all means. But never build this way for economic reasons only. See Chapter 18 and Fig. 18-2.

Chapter 24
Other Underground Homes

This is a chapter I've been looking forward to writing for some time now. The first and foremost reason is that it gives me a chance to talk about what someone else did, instead of constantly writing about what I did. Secondly, it gives me a space to express my opinion in reference to others involved in this unusual field of underground living.

THE PROBLEM WITH UNDERGROUND HOME SEMINARS

Now I finally get down to writing about what is irritating me. I have discussed with, spoken to, and addressed more people on this subject than 99 percent of the so-called experts on underground living. My book has been read by most people interested in the subject. My house has been on television, written about in newspapers and magazines and other books, but not once have I been asked to participate in one of these so-called seminars on underground homes sponsored by the well-known ecology oriented organizations. They claim to have all the experts, so that the people in attendance will benefit from their vast experiences. I suggest you take these seminars and meetings with a grain of salt, since they are not a real cross-section of the experts available.

Now before anyone jumps to the conclusion that this is only "sour grapes" talking, let me assure you that I don't need the exposure. I don't have the free time available and I don't work for free. The real point here, in all honesty, is that if you pay to attend a seminar on a subject, all the experts should be available to you, the potential underground home builder. It is not I who will miss anything; it is you. Now my thoughts are clear to be humble again and give real credit to other builders of underground homes.

THE WOODS' HOME

This interesting home is very unique in many ways, just as the owners and builders are. Freida and Earl Woods set about designing this home (Fig. 24-1) in 1976 after they purchased a lot in Boonelake Reservoir near Bluff City, Tennessee. They admitted to me that their first idea for an unusual home was not an underground home. They initially looked for an old yacht about 45 feet long. Their plans were to dry-dock it permanently on land by a lake and call it home. But Tennessee is not necessarily the big yacht center of the world, so they abandoned this boat idea.

Their second choice was going underground. After the normal hesitant moments and discussion, they began their plans (Fig. 24-2).

Earl was able to obtain valuable stress and strength information from a friend of a friend. This is what I've said so often. If you ask around, someone has the answers.

They began construction in April of 1977 doing most of the work themselves with the help of their son and daughter and their respective mates. The excavation and foundation were very similar to a conventional home, only deeper in the ground.

The Woods used a very unusual method of wall construction. I must admit I have never seen it before, because here in Maryland the codes would not permit this type of construction. Like I said before, different locales have different codes.

Fig. 24-1. The home of Freida and Earl Woods.

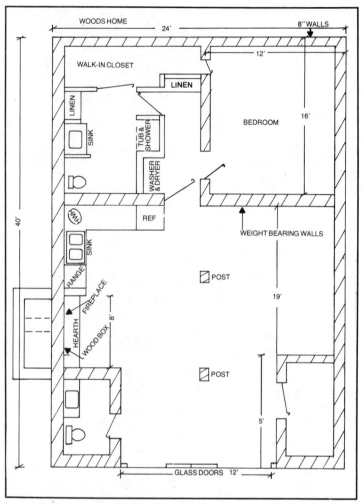

Fig. 24-2. Floor plan of the Woods' home.

The method they used was basic. They did not mortar the concrete blocks together. The blocks were only stacked in the same manner they would have been if mortar were used. Once the blocks were stacked satisfactorily, the exterior surface was covered with a layer of a cement-looking material using a trowel. This material worked like conventional stucco but was much stronger. The brand name is "Surwall" manufactured by W.R. Bonsal Company of

204

Lilesville, North Carolina. Earl Woods indicated it was very easy and fast to work with.

Once the walls were stacked like blocks and coated on the outside with "Surwall," the interior of this block was filled with concrete. Reinforcement rod was strategically inserted for strength. This is a very normal construction practice. Once the walls were solid and set, scaffolding was used to build a wooden structure to hold up the concrete roof as it was poured. Likewise, the roof slab has re-bar in it for strength.

Chapter 25
An Alternate
Method Of Construction

As with most things on this earth, there is more than one approach to underground home construction. My home, as with 99 percent of the other underground homes, is of concrete and block construction, but the one described on the following pages definitely is not. It is one of the few that is basically of wood construction. That wasn't a misprint—the primary wall and roof construction is wood. Please read and evaluate for yourself. If you do decide that this wood construction method is attrac've to you, I will remind you once again to consult with a q 'fied engineer on the subject of strength and permanency. The ne major point I want all potential builders to be cognizant of is that strength is the initial concern for either type of construction. However, concrete and block get stronger with age and moisture. Wood, on the other hand, is at its strongest the day it is placed. From that point on in time, status quo is the ultimate you can hope for. Mother Nature, in the form of insects, moisture and time, could play havoc with such construction.

I will discuss the home owned by J.A. Rinker from Traverse City, Michigan. The home was completed in the summer of 1977. The construction is basically wood, with 23 percent of the exterior walls exposed. If this type of dwelling is designed, built and maintained correctly, it should be stable for many years. Figure 25-1 shows the floor plan for the Rinker's home.

CONSTRUCTION STEPS FOR THE RINKER'S HOME

The following description of materials used in building the Rinkers' home is to be used as a guide and reference only. It is not intended to be a bill of materials for building an underground house.

As you might have suspected, the first step was to excavate a hole in the side of a hill and pour the footers and the 6-inch, reinforced floor slab over a standard, gravel-drain system and a 6-mil polyethylene vapor barrier. This is exactly as a conventional home would be built. Also, so far this is exactly like any underground home would be built. From this phase of construction on, it is definitely different than other underground homes.

The exterior walls are not block or concrete. They are wood, believe it or not. The 2 × 6s on 12-inch centers give strength and support to a layer of other material. Starting on the outside of the wall is special pressure-treated ¾-inch plywood, covered with two separate layers of 6-mil polyethylene. Between the 2 × 6 studs is 6 inches of fiber glass insulation. Underneath the interior ½-inch drywall is yet another layer of 6-mil polyethylene. Prior to pushing earth up against the outside, two layers of plastic seal, similar to thick paint, were applied to insure a watertight barrier. Refer to Figs. 25-2 and 25-3 for clarification. The roof was constructed in a very similar manner, except 2 × 12s were used, also on 12-inch centers.

First, the same ¾-inch treated plywood as in the walls was laid down on top of the 2 × 12s. On top of the plywood, there is ½-inch decking boards, and then five layers of asbestos felt with a layer of soft asphalt between each. Finally, a layer of 6-mil polyethylene is stretched over the

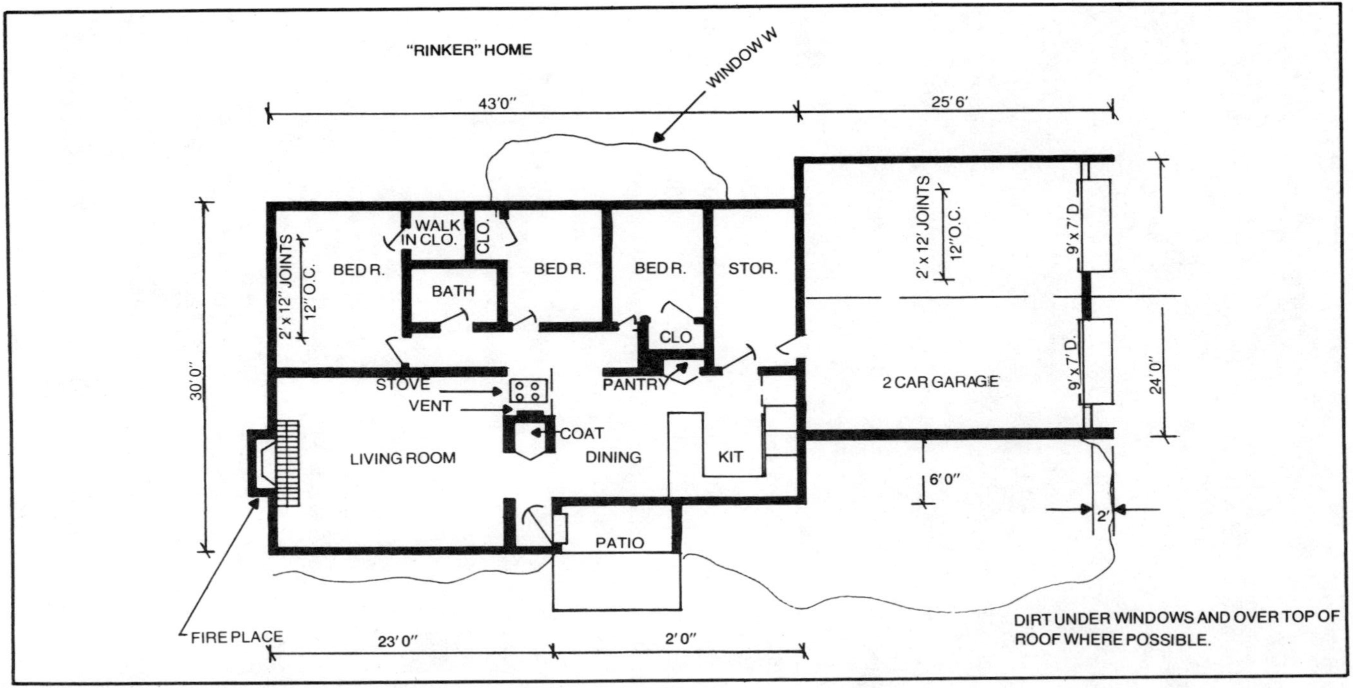
"RINKER" HOME
WINDOW W
43'0"
25' 6'
WALK IN CLO.
CLO.
BED R.
BED R.
BED R.
STOR.
BATH
2' x 12" JOINTS 12" O.C.
2' x 12' JOINTS 12" O.C.
9' x 7' D
9' x 7' D.
CLO
STOVE
VENT
PANTRY
COAT
2 CAR GARAGE
24'0"
30'0"
LIVING ROOM
DINING
KIT
6'0"
2'
PATIO
FIRE PLACE
23'0"
2'0"
DIRT UNDER WINDOWS AND OVER TOP OF ROOF WHERE POSSIBLE.

Fig. 25-1. Floor plan for the Rinker's home.

Fig. 25-2. View of the Rinker's house.

felt as a final moisture barrier. Now this roof is ready to be covered over with 8 inches of forest soil. At the same time, a 4-inch plastic drain pipe was laid around the perimeter of the roof as a safety factor.

Fig. 25-3. Another view of the Rinker's house.

OTHER FEATURES OF THE RINKER'S HOME

All interior wiring and plumbing is very similar to a conventional house, as is with most underground homes. The interior walls are all very standard construction, except for a load bearing block wall down the center of the house to hold up the 2 × 12s.

The Rinker's house is heated primarily with a fireplace. However, it does have electric baseboard heaters installed as a backup in case of extreme weather.

Since their house is located in Michigan, you had best believe that it has seen cold weather, as low as 40 degrees below zero. (That's cold, considering that in Maryland, the coldest my house has ever seen is about 1 degree above zero.)

This home is once again proof that a home doesn't have to be gigantic to be livable. This house has 1230 square feet of floor space, plus 612 square foot garage. The clever ideas involved here are proof positive that "Yankee" ingenuity is alive and well.

I can only remind you, as I did earlier, that this method of construction is less proven than others, but that doesn't make it any less an effective weapon against the oil companies. Read and investigate this method. It may be the one for you.

Chapter 26
The Add-On
Underground Home

This is an idea whose time has come, and it's a natural. My definition of an add-on underground house is self-explanatory. All you do is buy a small house or any old building (Fig. 26-1). The more picturesque, the better, in my opinion. This could really be any type of building from an old spring house (Fig. 26-2) to a small conventional farm building, or even a little bungalow already in use. The idea here is that you can easily use the small building as a starting point for building underground. More important, it can act as the entrance, very similar to the entrance chamber of many commercial caverns that people take tours through. This type of arrangement can be made very quaint and attractive by combining the old and the new. This idea, like all ideas, has drawbacks and advantages.

DISADVANTAGE

The only disadvantage I can think of is that these little buildings might be hard to find as it is. When you consider that you have to find them on a piece of property that is suitable for building underground, this task becomes even more difficult. However, if you look around you might get lucky (Figs. 26-3 and 26-4).

Fig. 26-1. An old building which could be renovated.

Fig. 26-2. An old spring house.

Fig. 26-3. Another old house.

Fig. 26-4. Another small building.

ADVANTAGES

The real advantage that you could realize from adding on to an existing building is, of all things, financing. In today's market (1980), mortgage money is difficult to come by. It is nearly impossible to find financing for unimproved land, which is what a vacant lot is. It is much easier to find financial help if a piece of ground has a driveway, well, septic system, electrical service and a mailbox. All of these things are essential anyway. It is usually cheaper to buy them already in use on a piece of land than it would be to install them at today's prices.

Buying a parcel of land with all services already in working order is a tremendous advantage to the man who is planning on building his own home as opposed to contracting it out. I'll tell you one big reason why, but first I'll point out that my suggestion most likely won't be taken well by neighbors or building inspectors. My suggestion is simple. Move a camper or an old trailer onto the property and hook into the existing utilities. Then you can continue working with relative ease. This gives you a place to keep food and drinks, to wash up and to take an occasional nap. I can't tell you what a drag it was to work all weekend without household conveniences.

All this might sound petty, but it isn't. Working outside in the sun is a back breaking job, and remember this book is written for the amateur builder. For any professional reading this book, I appreciate it, and I'm sure you know what I mean. You have a learned a long time ago that being comfortable is a better way to work than being uncomfortable.

See Figs. 26-5 and 26-6 for entrances to underground homes for which the owners will not give permission to expose their location because of privacy, and I will definitely respect their rights. All I'll say is one is in the Smokey Mountains and one is in the Northeast. But you get the idea as to how nice an entrance an existing little building makes for an underground home.

Fig. 26-5. This homeowner wants privacy.

Fig. 26-6. A little building makes a nice entrance for an underground home.

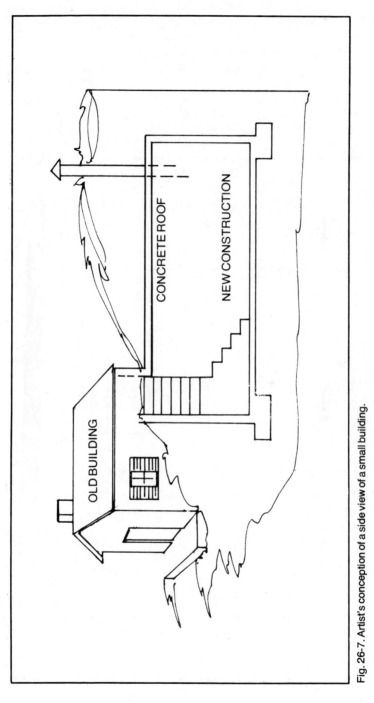

Fig. 26-7. Artist's conception of a side view of a small building.

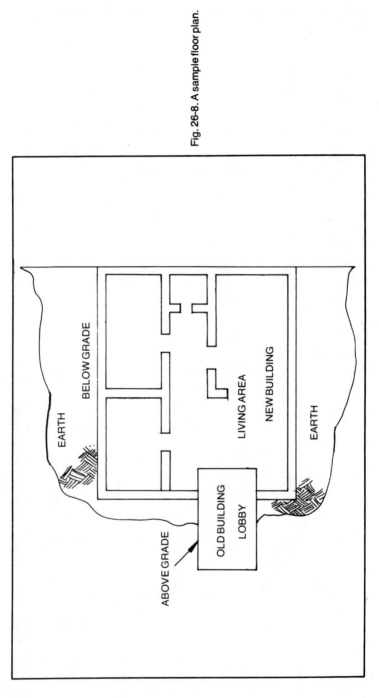

Fig. 26-8. A sample floor plan.

I suggest you give the idea a lot of thought. I plan to design a few of these homes for people who have contacted me, and I must say that I would give it serious thought if I had it to do over again. If one type of underground construction is going to catch on with the general public, I feel it will be in this category of "add-on" underground homes. It's a natural, especially if anyone already owns their own property and a small building with utilities.

THEORETICAL DESIGN

The following idea is just that, but it could apply to any of the small buildings shown in this chapter. See Fig. 26-7 for a basic artist's conception of the side view. See Fig. 26-8 for a sample floor plan. Remember, this is only an example of the many different possibilities that exist. Let your imagination take over for a while.

If a problem exists, it will be sealing the joints between the old building and the new construction. Of course, the excavator had better be careful with the backhoe and bulldozer or he will destroy the existing building. At least the foundation could be weakened.

Chapter 27
Sound Level In
Underground Homes

Yet another difference you will find by living underground is that the sound level is changed drastically. In my house it is noticeable from room to room, because each interior wall is block and Sheetrock. These materials absorb sound well, and each room has good quality carpet on the floors. However, the greatest sound transfer change is from the exterior to the interior.

COMPARISON WITH THE
SOUND LEVEL IN CONVENTIONAL HOMES

Since everyone is familiar with standard home living, it is difficult to explain in writing what the difference in sound level really is. But I'll try. Conventional homes allow the transmission of normal everyday sounds such as passing cars, bird calls, dogs barking and even the wind blowing. These sounds are so accepted to the human ear that you will think it is really quiet in the house, even though those sounds I just mentioned are prevalent. The true absence of sound is almost never heard under normal living conditions—that is, until you build underground.

Fig. 27 - 1. The skylight is in place over the garden.

You can imagine what it is like to not hear traffic or any of the accepted noises previously mentioned. This is not an imagined difference; it is for real. If it wasn't for the skylight over the garden area, we wouldn't know if it was raining or snowing without going outside. As it is, we have to go into the garden to check the weather when most people just look out the window. If this bothers anyone, I suggest your design be different than mine.

CHIMNEY NOISE

There is one clue that it is raining or the wind is blowing outside. At times, if conditions are right, the rain hits the chimney. Since the chimney is steel pipe instead of masonry construction, we sometimes get strange noises coming out of the stove. But I still wouldn't trade my setup for any other house. If these are the only disadvantages I have in life, I'll learn to live with them.

Chapter 28
Exterior Walls

Once the building is covered with earth as much as it is going to be, you want to put facing on the block work. This could be any one of the accepted materials such as aluminum siding, cedar shingles, rock, stone or brick. In my case, I decided to use my own concoction of cement, sand, gravel and anything else that bonded together in concrete to mix up a type of stucco. See Figs. 28-1 through 28-3.

PAINTING WALLS

If you have ever mixed concrete, batch after batch, you know that the coloration of each batch is different from the next. I solved this very easily by mixing a gray latex paint very thin (about 10 parts water to 1 part paint) to form a dye. I used a small compressor and spray gun to lightly spray the complete structure. No one could tell that the surface had been painted. It was a natural match. This painting has another advantage that turns out to be great. Every spring I give it another light coat. It only takes about two hours to do all exposed walls. Each spring the house looks brand new. I love the idea.

Fig. 28-1. View of an exterior wall.

Fig. 28-2. Make up your own design.

230

Fig. 28-3. A painted wall.

GARDEN WALL

One wall in my garden has old roof slate glued on just as you see it in Fig. 28-4. Use the same glue that you used elsewhere in the house construction. Most people think it is

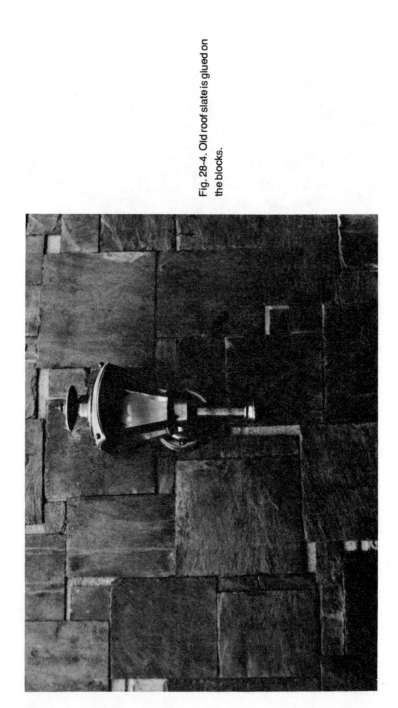

Fig. 28-4. Old roof slate is glued on the blocks.

unique. Examine the pictures in this book closely, maybe even with a reading glass. You'll pick up a lot of useful information.

Chapter 29
Taxes

Any time taxes are mentioned in reference to underground homes, it is usually real estate taxes that are up for discussion. The term *real estate tax* is self-explanatory. It is the tax that the local government imposes on your real property (land and building) to cover the cost of roads, schools, police and fire protection, and to pay all the county services such as the health department, legal clinics and many others. Naturally, the more services required by a community, the higher the tax will be. It doesn't take much to figure out that the more people in an area, the more services required. Thus, the area will have a higher tax rate. If you think this isn't true, check out the tax rate per $100 of property value in Baltimore, Maryland. Then check out the tax rate in central Arkansas.

Real estate taxes upset the old axiom that you get what you pay for. In most cases, the majority of the taxpayers get very little for their input and a few non-taxpayers reap the benefits.

DETERMINING THE TAX BILL

Your annual tax bill is probably computed on a formula similar to the one I'll use in the following examples. Your local tax assessor estimates the value of your land and buildings. He pulls this figure out of thin air using only figures from recent sales of similar properties as a guide.

If you build a 2,000 square foot, three bedroom, brick ranch house on a half-acre lot, and a house very similar to yours, a half mile down the road, sold last week for $50,000, then you can expect to receive a tax bill based on $50,000 as the value of your home. Just because you built it from free material or free labor has nothing to do with the actual value when finished. Square feet of living space, type of construction and location all determine what the value of your property will be.

This brings me back to the original point. Underground home taxes are no different than those for a conventional home because of the unwritten rules I just mentioned—that is, square feet of living space, method of construction and location. It makes no difference that a home is 10 feet underground or not. For that matter, it could be 10 feet in the air. It is still living space in a specific location.

WHAT AFFECTS TAX RATES?

In most cases, very little affects tax rate. The rate is the figure in dollars and cents (example $2.80) that is preset by the local government. It is a simple calculation of how much money you pay per $100 of property value. If your local tax rate is as my example, $2.80, per $100 value, and the tax assessor says that your property is worth $10,000 it is easy to figure your tax bill (100 × $2.80 = $280).

ASSESSMENT

Tax rate is one thing. Assessment is another. Assessment is the total value of your property (building and land). The tax assessor, as I said before, estimates your value, based on past records, experience and neighboring properties.

As opposed to tax rate where virtually nothing can be changed, the assessed value can fluctuate a little now and then. For example, swimming pools, outside sheds and buildings, paved driveways, patios, outside fireplaces and landscaping all are taken into consideration when your property is appraised. So your house could be exactly like your neighbor's, but you could have all the items I just mentioned and, of course, your assessment will be proportionately higher. If you want to keep your tax bill down, keep your home more basic. Frills are nice, but you pay for them again and again through taxes.

FIGURING THE TAX RATE

There just aren't many arguments against real estate taxes related to underground homes, unless you use the one that I did. Underground homes have no *established* resale value. So it is unfair to judge an underground home with a neighboring house of equal square footage. Everyone agrees that the resale value is not a matter of record, but the tax assessor couldn't care less about that. Don't put much stock in this argument. It may or may not be worth a few dollars.

If you wonder how that example $2.80 figure would be derived, I'll devote a few sentences to a simple explanation. Your local government (county) estimates how much money they will need to keep the county going for the next year. This includes salaries, equipment, payoffs and so on. They know what the value of all the land is in their tax district from past years, plus new building permits that were issued.

From here on, it is basic division. Divide what the government needs to run the county (example 28 million) for a year *into* what the total value of all the taxable real estate is. The value of all the land is worth about $100 million. Do the division and you get the $2.80 figure to use as a tax rate. It isn't that simple, but I'm sure you get the picture.

The only variation to this rate could be deductions for senior citizens in some cases, non-profit organizations, and producing farm or commercial establishments. These are a few possibilities where the tax rate per $100 value could vary. Naturally, each locale is different. But don't count on falling into one of these categories. The majority of the homeowners fall under the basic established rate.

Chapter 30
Resale Value

The biggest comeback the bank has for not lending money to you for building underground is that nobody knows what the *resale value* is. Therefore, the value is minimal. I can in truth say that I couldn't find anyone who has bought or sold an underground house to try and get a firsthand example of selling cost compared to original building cost. So I'll have to revert back to my own experiences.

VISITORS AND INQUIRIES

Ever since the day we moved into this house, we have had visitors and inquiries from around the country. I don't recall how many people have offered to buy it when I'm ready to sell, but there have been quite a few. And these are totally unsolicited inquiries—no "for sale" sign, nothing. People come right out and ask if the house is for sale. Therefore, I draw the conclusion that someone would be willing to put their money where their mouth is, even if you consider that most people were only curious. "It only takes one to sell," as the saying goes in the real estate business.

Fig. 30-1. Everyone will want to visit your underground home.

240

REAL ESTATE AGENTS

This leads me to what real estate agents have said. I have had many agents call me inquiring if I'm interested in selling. One agent with offices nationwide said he would advertise the house in national newspapers. Believe me, they wouldn't offer to do this if they didn't think they could get top dollar. I'm very confident that the resale value of my house is greater than that of a conventional house. Here's a note of interest about selling an underground house. Don't forget that if it is on the market, every curiosity seeker within gunshot will pretend to be interested just to see the inside. Work out a plan with a real estate agent that screens out persons who can't pass a financial statement check. It will save wear and tear on your house and nerves.

Chapter 31
Abandoned
Underground Structures

Just as in any other endeavor, sometimes something goes haywire with underground buildings. The building in Fig. 21-1 is a poured concrete structure built many years ago. The exact date is unknown, but the building is probably 80 years old or so.

STORE AND SUPPLY STATION

This building was used as a store and supply station near a railroad. I looked into the building and examined the structure. I found numerous cracks, but nothing had fallen down. The reasons for the cracks were two fold. First, freezing and thawing occurred year after year, because the door and windows were mostly missing, and, second, no humans lived there.

AMMUNITION STORAGE ROOM

The structure in Fig. 31-2 seems to resemble a modern day bridge support, when in fact it was constructed at the same time the store building was built. The best anyone can figure is that this building was used as an ammunition storage

Fig. 31-1. A poured concrete structure.

Fig. 31-2. This structure might have been used as an ammunition storage room.

room. Needless to say, it worked, or at least if the ammunition went off at any time it didn't damage the building. Note that underground structures always were around and always will be. They are just not as obvious as buildings stuck on top of the ground.

Chapter 32
Unconventional
Underground Home Design

One of the side benefits that has come to me by way of my underground home building project was the writing of books on the subject. One of the side benefits of writing books is speaking to organizations such as schools and community groups.

At these speaking engagements, I always have a question and answer period for the audience. I really like these discussions because of the unique and clever ideas I hear about. The following are sketches and basic details of a few of the more realistic ideas that have been brought to my attention. I admit that I probably wouldn't endorse any of them as being practical, but I can remember when my house wasn't considered practical, either. Who knows what the future of underground homes holds?

ADVANTAGES AND DISADVANT-
AGES WITH THE PYRAMID DESIGN

The *pyramid* design is a combination of steel and concrete (Fig. 32-1). This design is only on the drawing board as far as I know. So any information I give on this design has not been proven by time (Fig. 32-2).

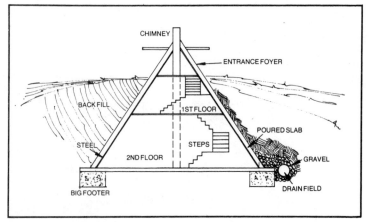

Fig. 32-1. A pyramid design.

If there is one advantage to this design, it is that construction should progress very quickly. Steel erection goes pretty fast once a welder and crane are in place. Once the steel is welded to form a stable structure it will also be very easy to make an entrance to the ground level. Once you get past these two advantages (if they really are), there aren't many more.

The biggest single disadvantage is all that steel underground. You must realize that unless steel is coated correctly, it will rust away very quickly. Steel can be chemically coated to prevent rusting, but this procedure is

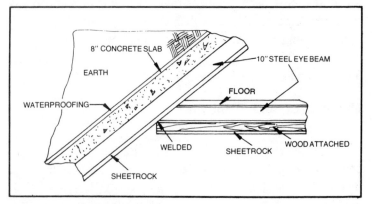

Fig. 32-2. A lot of welding is required.

expensive. Once covered with earth, no one knows if the steel is rusting or not.

EXPENSE

If you think coating steel to prevent it from rusting is expensive, wait until you price the steel beams themselves. On top of the initial cost, you need a crane to put steel in place and a welder to make it stay put. Once you get the estimates on the steel construction phase of this design, I think you see why I don't recommend building with steel supports, such as this design would require. It would take a real professional to successfully build this one.

WATERPROOFING

Waterproofing would be similar to any other underground design. The soil is the critical subject, not the structure design.

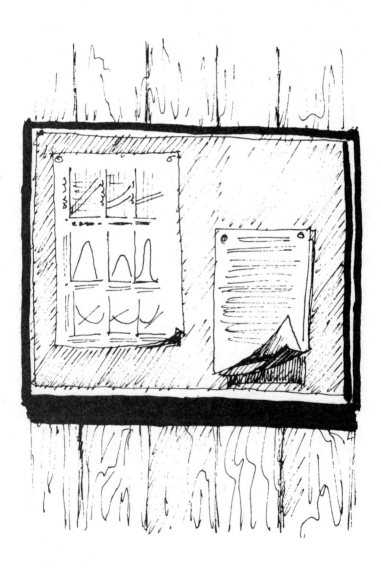

Chapter 33
Underground Home Publicity

There is a fact that you must face when getting involved in any project as big as building an underground house. As with most unusual projects, many people will be interested. It seems that with the energy shortage, underground homes are surely a subject to get attention. Of course, there are many types of attention that you can get—negative, positive, private and public.

POSITIVE ATTENTION

Fortunately for you as the home builder, the people that give you publicity, whether it be by their request or yours, are fairly easy to categorize. As you read on, you will see what I mean.

I have found one fact to be 99 per cent true. The groups of individuals who personally ask to see your house are almost always friendly and not likely to cause you any problem. You very rarely will have a person ask to see your house and then bad mouth it behind your back. If you build your underground house correctly, visitors will go away impressed. They will also become your best moral supporters. You will be surprised by the number of friendships you make that will continue after the house building has been completed. This is reason enough to build an underground house—we all need all the friends we can get.

NEGATIVE ATTENTION

Just as the private citizens who ask to see your house are 99 percent friendly, you can rest assured that 99 percent of the public or regulatory personnel who see your house will have a negative opinion. In all fairness to these people, they usually form their opinions through the eyes of their specific jobs—zoning, health, fire, insurance or building inspectors. Don't be surprised if the dog catcher even gets into the act! Let me explain what I mean be seeing an underground house through their job titles as opposed to their personal interest.

In the course of my complicated maneuverings with the county and state officials, I had one inspector who gave me real problems as he acted in the manner and capacity of an inspector. As a matter of fact, he was downright uncooperative. However, a week after the inspection, the same inspector contacted me and wanted to know if he could show my house to his family. This time everyone was as friendly as could be. Once back on the job, he reverted to his old self.

PUBLIC ATTENTION

This is the most critical type of attention you will receive. It can make your adaption into the community smooth, or if the attention the newspaper, radio and television give you is negative, the neighborhood will be convinced that your house is a black spot in the community. At this point, I will put your mind at ease and tell you to relax. When a newspaper reporter or television station contact you for an interview, they are almost always forward-thinking, intelligent individuals who like to see people doing individualistic projects, especially saving energy. The time is right. These reporters can be your best allies if you get into any real hassles with officials. Reporters, by nature, will see that no one pulls the wool over anyone's eyes simply by constant exposure in the media.

I do, however, have some good advice for you as you prepare for your interviews with the media, be it television or newspapers. Know what you are talking about; don't make dumb, specific statements. First of all, they only want general information that the public can relate to. For example, if a reporter asks you how strong your concrete roof is, don't answer that it will hold up exactly 795 pounds per square foot. First of all, no one knows exactly how strong your concrete is and secondly, someone will begin to question your judgment. Whatever you do, don't pin yourself into a corner by broadcasting specific facts about your house that the public does not need to know.

All the reporters I have talked to have been friendly, so be sure to act accordingly. They are great people to have on your side to spread the word about the good points of your underground house.

The one reservation you must have around reporters is to be exact as to what you want to be *on the record* or *off the record*. This, of course, has nothing to do with underground homes, but since you probably are not familiar with inter-view procedures, I will only tell you that once a statement is made, it can not be retracted. However, a reporter will not report anything preceded by the words, "This is off the record."

PRIVATE ATTENTION

This is the same as acceptance by or rejection by the neighborhood. The less negative attention you receive publicly, the less private attention you are likely to receive. Accept the fact that building an underground house is an attention-getting project. Use it to your advantage and do not let it cause you a problem.

Glossary

adhesive strength: The quality of the bond that mortar has for holding two masonry units together.

aggregate: Various hard, inert materials such as sand, gravel or pebbles in various size fragments mixed with cementing material to form concrete, mortar or plaster.

amalgamation: A mixed blend or combination of materials such as lime, cement, sand and water.

atrium: "Garden" area covered by a skylight. It should have an earth floor.

basement: The substructure of a building that is wholly or partially below ground level.

bearing wall: A partition which supports the weight of a structure.

block: A solid unit of masonry material formed in a uniform size.

cement: A powdered mixture of alumina, silica, lime, iron oxide and magnesia burned together in a kiln and finely pulverized. When it is mixed with water, it forms a plastic mass that hardens by chemical combination.

chimney: A vertical masonry structure for carrying off smoke.

dome: curved surface skylight.

earth-sheltered home: Any structure that uses the earth to assist in insulation.

energy efficient: Term referring to anything that requires less fuel than it did before the first major oil embargo.

footing: Foundation or bottom unit of a wall.

foundation: The portion of a structure, usually below ground level, that supports a wall or other structure.

graded aggregate: Aggregate with various types of sand and gravel.

grade level: Average level of the land as it is naturally before man has disturbed it.

hardwood: The close-grained wood from broad-leaved trees such as oak or maple.

joint: The space between two adjacent masonry units bonded by mortar.

lean mortar: Mortar that is deficient in bonding material.

masonry: Walls built by a mason using brick, stone, tile or similar materials.

mortar: Mixture of lime, cement and water used to bond masonry.

mortar board: A small, lightweight board on which masons temporarily place mortar.

natural draw: Warm air going up through the chimney drawing outside air inside.

neat cement: Cement and water without aggregate.

plenum: Chamber that connects the furnace to the duct system.

portland cement: Hydraulic cement made by burning a mixture of limestone and clay in a kiln.

re-bar: A steel reinforcing bar embedded in concrete construction to add strength.

retaining wall: A wall built to prevent the soil at an embankment or cut from sliding.

single pour: A continuous flow of wet concrete until a specific job is complete, regardless of quantity.

skylight: Any covered structure that lets light into a building.

slab: Concrete floor placed directly on earth or a gravel base and usually about 4 inches thick.

solar heat: Heat from the sun.

sound transmission: Passage of sound through a construction assembly.

subfloor: Usually, plywood sheets that are nailed directly to the floor joists and that receive the finish flooring.

underground home: Any building with at least 3 feet of earth over 100 percent of the roof and at least 70 percent of wall surfaces covered.

Appendix A
Weights and
Specific Gravities

WEIGHTS AND SPECIFIC GRAVITIES

Substance	Weight Lb. per Cu. Ft.	Specific Gravity
METALS, ALLOYS, ORES		
Aluminum, cast, hammered	165	2.55–2.75
Brass, cast, rolled	534	8.4–8.7
Bronze, 7.9 to 14%, Sn	509	7.4–8.9
Bronze, aluminum	481	7.7
Copper, cast, rolled	556	8.8–9.0
Copper ore, pyrites	262	4.1–4.3
Gold, cast, hammered	1205	19.25–19.3
Iron, cast, pig	450	7.2
Iron, wrought	485	7.6–7.9
Iron, spiegel-eisen	468	7.5
Iron, ferro-silicon	437	6.7–7.3
Iron ore, hematite	325	5.2
Iron ore, hematite in bank	160–180	
Iron ore, hematite loose	130–160	
Iron ore, limonite	237	3.6–4.0
Iron ore, magnetite	315	4.9–5.2
Iron slag	172	2.5–3.0
Lead	710	11.37
Lead ore, galena	465	7.3–7.6
Magnesium, alloys	112	1.74–1.83
Manganese	475	7.2–8.0
Manganese ore, pyrolusite	259	3.7–4.6
Mercury	849	13.6
Monel Metal	556	8.8–9.0
Nickel	565	8.9–9.2

Substance	Weight Lb. per Cu. Ft.	Specific Gravity
TIMBER, U.S. SEASONED		
Moisture Content by Weight:		
Seasoned timber 15 to 20%		
Green timber up to 50%		
Ash, white, red	40	0.60–0.62
Cedar, white, red	22	0.32–0.38
Chestnut	41	0.66
Cypress	30	0.48
Fir, Douglas spruce	32	0.51
Fir, eastern	25	0.40
Elm, white	45	0.72
Hemlock	29	0.42–0.52
Hickory	49	0.74–0.84
Locust	46	0.73
Maple, hard	43	0.68
Maple, white	33	0.53
Oak, chestnut	54	0.86
Oak, live	59	0.95
Oak, red, black	41	0.65
Oak, white	46	0.74
Pine, Oregon	32	0.51
Pine, red	30	0.48
Pine, white	26	0.41
Pine, yellow, long-leaf	44	0.70
Pine, yellow, short-leaf	38	0.61
Poplar	30	0.48

Material		
Platinum, cast, hammered	1330	21.1–21.5
Silver, cast, hammered	656	10.4–10.6
Steel, rolled	490	7.85
Tin, cast, hammered	459	7.2–7.5
Tin ore, cassiterite	418	6.4–7.0
Zinc, cast, rolled	440	6.9–7.2
Zinc ore, blende	253	3.9–4.2

VARIOUS SOLIDS

Material			
Cereals, oats	bulk	32	
Cereals, barley	bulk	39	
Cereals, corn, rye	bulk	48	
Cereals, wheat	bulk	48	
Hay and Straw	bales	20	
Cotton, Flax, Hemp		93	1.47–1.50
Fats		58	0.90–0.97
Flour, loose		28	0.40–0.50
Flour, pressed		47	0.70–0.80
Glass, common		156	2.40–2.60
Glass, plate or crown		161	2.45–2.72
Glass, crystal		184	2.90–3.00
Leather		59	0.86–1.02
Paper		58	0.70–1.15
Potatoes, piled		42	
Rubber, caoutchouc		59	0.96–0.9.
Rubber goods		94	1.0–2.0
Salt, granulated, piled		48	
Saltpeter		67	
Starch		96	1.53
Sulphur		125	1.93–2.07
Wool		82	1.32

Material		
Redwood, California	26	0.42
Spruce, white, black	27	0.04–0.40
Walnut, black	38	0.61
Walnut, white	26	0.41

VARIOUS LIQUIDS

Material		
Alcohol, 100%	49	0.79
Acids, muriatic 40%	75	1.20
Acids, nitric 91%	94	1.50
Acids, sulphuric 87%	112	1.80
Lye, soda 66%	106	1.70
Oils, vegetable	58	0.91–0.94
Oils, mineral, lubricants	57	0.90–0.93
Water, 4°C. max. density	62.428	1.0
Water, 100°C.	59.830	0.9584
Water, ice	56	0.88–0.92
Water, snow, fresh fallen	8	.125
Water, sea water	64	1.02-1.03

GASES

Material		
Air. 0°C. 760 mm.	.08071	1.0
Ammonia	0.478	0.5920
Carbon dioxide	.1234	1.5291
Carbon monoxide	.0781	0.9673
Gas, illuminating	.028–.036	0.35–0.45
Gas, natural	.038–.039	0.47–0.48
Hydrogen	.00559	0.0693
Nitrogen	.0784	0.9714
Oxygen	.0892	1.1056

WEIGHTS AND SPECIFIC GRAVITIES

Substance	Weight Lb. per Cu. Ft.	Weight Lb. per Cu. Ft.
ASHLAR MASONRY		
Granite, syenite, gneiss	165	2.3–3.0
Limestone, marble	160	2.3–2.8
Sandstone, bluestone	140	2.1–2.4
MORTAR RUBBLE MASONRY.		
Granite, syenite, gneiss	155	2.2–2.8
Limestone, marble	150	2.2–2.6
Sandstone, bluestone	130	2.0–2.2
DRY RUBBLE MASONRY		
Granite, syenite, gneiss	130	1.9–2.3
Limestone, marble	125	1.9–2.1
Sandstone, bluestone	110	1.8–1.9
BRICK MASONRY		
Pressed brick	140	2.2–2.3
Common brick	120	1.8–2.0
Soft brick	100	1.5–.7
CONCRETE MASONRY		
Cement, stone, sand	144	2.2–2.4

Substance	Weight Lb. per Cu Ft.	Specific Gravity
EXCAVATIONS IN WATER		
Sand or gravel	60	
Sand or gravel and clay	65	
Clay	80	
River mud	90	
Soil	70	
Stone riprap	65	
MINERALS		
Asbestos	153	
Barytes	281	
Basalt	184	
Bauxite	159	
Borax	109	
Chalk	137	
Clay, marl	137	
Dolomite	181	
Feldspar, orthoclase	159	
Gneiss, serpentine	159	
Granite, syenite	175	
Greenstone, trap	187	
Gypsum, alabaster	159	
Hornblende	187	
Limestone, marble	165	
Magnesite	187	
Phosphate rock, apatite	200	
Porphyry	172	
Pumice, natural	40	
Quartz, flint	165	
Sandstone, bluestone	147	

VARIOUS BUILDING MATERIALS

Ashes, cinders	40–45	
Cement, portland, loose	90	
Cement, portland, set	183	2.7–3.2
Lime, gypsum, loose	53–64	1.4–1.9
Mortar, set	103	
Slags, bank slag	67–72	
Slags, bank screenings	98–117	
Slags, machine slag	96	
Slags, slag sand	49–55	

EARTH, ETC., EXCAVATED

Clay, dry	63
Clay, damp, plastic	110
Clay and gravel, dry	100
Earth, dry, loose	76
Earth, dry, packed	95
Earth, moist, loose	78
Earth, moist, packed	96
Earth, mud, flowing	108
Earth, mud, packed	115
Riprap, limestone	80–85
Riprap, sandstone	90
Riprap, shale	105
Sand, gravel, dry, loose	90–105
Sand, gravel, dry, packed	100–120
Sand, gravel, dry, wet	118–120

STONE, QUARRIED, PILED

Basalt, granite, gneiss	96
Limestone, marble, quartz	95
Sandstone	82
Shale	92
Greenstone, hornblende	107

BITUMINOUS SUBSTANCES

Asphaltum	81
Coal, anthracite	97
Coal, bituminous	84
Coal, lignite	78
Coal, peat, turf, dry	47
Coal, charcoal, pine	23
Coal, charcoal, oak	33
Coal, coke	75
Graphite	131
Paraffine	56
Petroleum	54
Petroleum, refined	50
Petroleum, benzine	46
Petroleum, gasoline	42
Pitch	69
Tar, bituminous	75

COAL AND COKE, PILED

Coal, anthracite	47–58
Coao, bituminous, lignite	40–54
Coal, peat, turf	20–26
Coal, charcoal	10–14
Coal, coke	23–32

The specific gravities of solids and liquids refer to water at 4°C., those of gases to air at 0°C, and 760 mm. pressure. The weights per cubic foot are derived from average specific gravities, except where stated that weights are for bulk, heaped or loose material, etc.

Appendix B
Underground
House Statistics

- HOUSE SIZE—40′ × 90′ (3600 SQUARE FEET)
- FIFTEEN ROOMS, PLUS GARAGE
- OVER 7500 CONCRETE BLOCK USED
- OVER 250 CUBIC YARDS OF CONCRETE
- APPROXIMATELY TEN TONS OF STEEL USED
- OVER 800 TONS OF DIRT ON ROOF
- WOOD HEAT STOVE ONLY

Appendix C
Weight of Basic Materials
Used in Underground Construction

Weight of Basic Materials Used In Underground Construction		
SANDY SOIL	1 CUBIC FT.	65 LB
MUD	1 CUBIC FT.	90 LB
WATER	1 CUBIC FT.	62½ LB
CONCRETE	1 CUBIC FT.	140 LB
4' × 8' SHEET ¼"	ACRYLIC PLASTIC	47 LB
4' × 8' SHEET ½"	SHEET ROCK (GYPSUM BOARD)	65 LB
4' × 8' SHEET ½"	PLYWOOD	55 LB
12" × 8" × 8"	CONCRETE BLOCK	54 LB
6" × 8" x8"	CONCRETE BLOCK	38 LB

Appendix D
Slump Test

The slump test is used to measure the consistency of the concrete. The test is made by using a *slump cone*; the cone is made of No. 16 gauge galvanized metal with the base 8 inches in diameter, the top 4 inches in diameter, and the height 12 inches. The base and the top are open and parallel to each other and at right angles to the axis of the cone. A tamping rod ⅝ inch in diameter and 24 inches long is also needed. The tamping rod should be smooth and bullet pointed (not a piece of rebar).

Samples of concrete for test specimens should be taken at the mixer or, in the case of ready-mixed concrete, from the transportation vehicle during discharge. The sample of concrete from which test specimens are made will be representative of the entire batch. Such samples should be obtained by repeatedly passing a scoop or pail through the discharging stream of concrete, starting the sampling operation at the beginning of discharge, and repeating the operation until the entire batch is discharged. The sample being obtained should be transported to the testing site. To counteract segregation, the concrete should be mixed with a shovel until the concrete is uniform in appearance. The location in the work of the batch of concrete being sampled should be noted for future reference. In the case of paving concrete, samples may be taken from the batch immediately after depositing on the subgrade. At least five samples should be taken from different portions of the pile and these samples should be thoroughly mixed to form the test specimen.

The cone should be dampened and placed on a flat, moist nonabsorbent surface. From the sample of concrete obtained, the cone should immediately be filled in three layers, each approximately one-third the volume of the cone. In placing each scoopful of concrete the scoop should be moved around the top edge of the cone as the concrete slides from it, in order to ensure symmetrical distribution of concrete within the cone. Each layer should be *rodded in* with 25 strokes. The strokes should be distributed uniformly over the cross section of the cone and should penetrate into the underlying layer. The bottom layer should be rodded throughout its depth.

When the cone has been filed to a little more than full, strike off the excess concrete, flush with the top, with a straightedge. The cone should be immediately removed from the concrete by raising it carefully in a vertical direction. The slump should then be measured to the center of the slump immediately by determining the difference between the height of the cone and the height at the vertical axis of the specimen as shown in Fig. D-1.

The consistency should be recorded in terms of inches of subsidence of the specimen during the test, which is called slump. Slump equals 12 inches of height after subsidence.

After the slump measurement is completed, the side of the mix should be tapped gently with the tamping rod. The behavior of the concrete under this treatment is a valuable indication of the cohesivesness, workability, and placeability

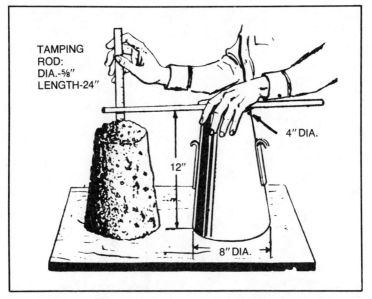

Fig. D-1. Measurement of slumps.

of the mix. A well-proportioned workable mix will gradually slump to lower elevations and retain its original identity, while a poor mix will crumble, segregate, and fall apart.

Appendix E
Cement Statistics

Table E-1. Age Compression Strength
Relationship for Types I and III Air-Entrained Portland Cement.

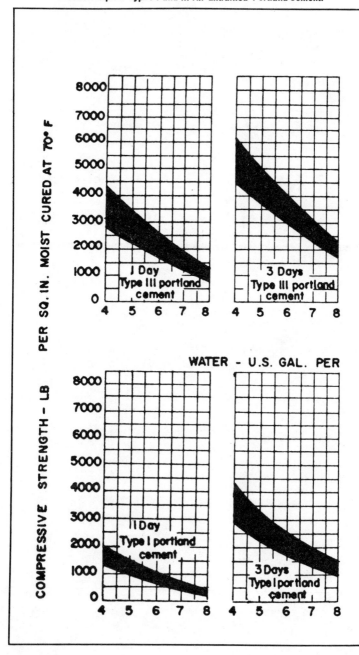

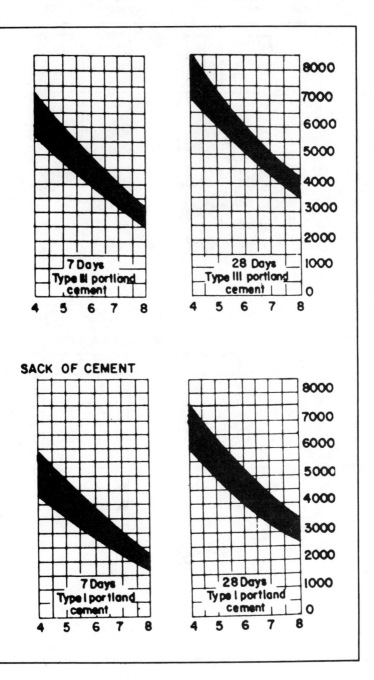

SACK OF CEMENT

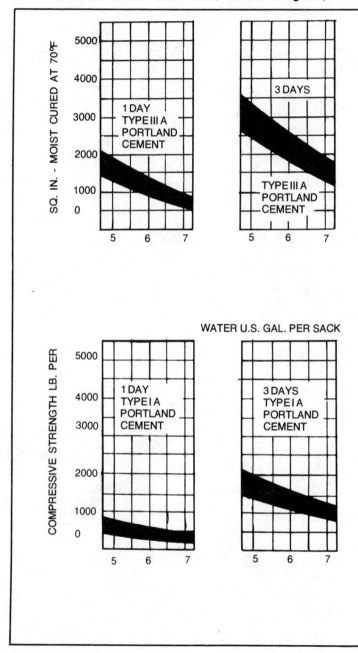

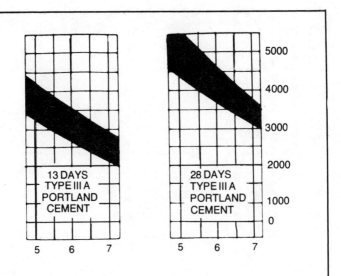

OF CEMENT

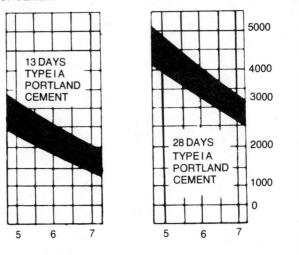

Table E-2. Suggested Trial Mixes for Non-Entrained Concrete of Medium Consistency with 3- to 4-Inch Slump.

Water-cement ratio Gal per sack	Maximum size of aggregate inches	Air content (entrapped air) per cent	Water gal per cu yd of concrete	Cement sacks per cu yd. of concrete	With fine sand fineness modulus = 2.50		
					Fine aggregate -per cent of total aggregate	Fine aggregate lb per cu yd of concrete	Coarse aggregate lb per cu yd of concrete
4.5	3/8	3	46	10.3	50	1240	1260
	1/2	2.5	44	9.8	42	1100	1520
	3/4	2	41	9.1	35	960	1800
	1	1.5	39	8.7	32	910	1940
	1½	1	36	8.0	29	880	2110
5.0	3/8	3	46	9.2	51	1330	1260
	1/2	2.5	44	8.8	44	1180	1520
	3/4	2	41	8.2	37	1040	1800
	1	1.5	39	7.8	34	990	1940
	1½	1	36	7.2	31	960	2110
5.5	3/8	3	46	8.4	52	1390	1260
	1/2	2.5	44	8.0	45	1240	1520
	3/4	2	41	7.5	38	1090	1800
	1	1.5	39	7.1	35	1040	1940
	1½	1	36	6.5	32	1000	2110

6.0	3/8	3	46	7.7	53	1440	1260
	1/2	2.5	44	7.3	46	1290	1520
	3/4	2	41	6.8	39	1130	1800
	1	1.5	39	6.5	36	1080	1940
	1½	1	36	6.0	33	1040	2110
6.5	3/8	3	46	7.1	54	1480	1260
	1/2	2.5	44	6.8	46	1320	1520
	3/4	2	41	6.3	39	1190	1800
	1	1.5	39	6.0	37	1120	1940
	1½	1	36	5.5	34	1070	2110
7.0	3/8	3	46	6.6	55	1520	1260
	1/2	2.5	44	6.3	47	1360	1520
	3/4	2	41	5.9	40	1200	1800
	1	1.5	39	5.6	37	1150	1940
	1½	1	36	5.1	34	1100	2110
7.5	3/8	3	46	6.1	55	1560	1260
	1/2	2.5	44	5.9	48	1400	1520
	3/4	2	41	5.5	41	1240	1800
	1	1.5	39	5.2	38	1190	1940
	1½	1	36	4.8	35	1130	2110
8.0	3/8	3	46	5.7	56	1600	1260
	1/2	2.5	44	5.5	48	1440	1520
	3/4	2	41	5.1	42	1280	1800
	1	1.5	39	4.9	39	1220	1940
	1½	1	36	4.5	35	1160	2110

*See footnote at end of table.

(Continued on next page)

279

Table E-2. Suggested Trial Mixes for Non-Entrained Concrete of Medium Consistency with 3- to 4-Inch Slump. (Continued from Page 279).

Water-cement ratio Gal per sack	With average sand—fineness modulus = 2.75			With coarse sand—fineness modulus = 2.90		
	Fine aggregate percent of total aggregate	Fine aggregate lb per cu yd of concrete	Coarse aggregate lb per cu yd of concrete	Fine aggregate percent of total aggregate	Fine aggregate lb per cu yd of concrete	Coarse aggregate lb per cu yd of concrete
4.5	52	1310	1190	54	1350	1150
	45	1170	1450	47	1220	1400
	37	1030	1730	39	1080	1680
	34	980	1870	36	1020	1830
	32	960	2030	33	1000	1990
5.0	54	1400	1190	56	1440	1150
	46	1250	1450	48	1300	1400
	39	1110	1730	41	1160	1680
	36	1060	1870	38	1100	1830
	34	1040	2030	35	1080	1990
5.5	55	1460	1190	57	1500	1150
	47	1310	1450	49	1360	1400
	40	1160	1730	42	1210	1680
	37	1110	1870	39	1150	1830
	35	1080	2030	36	1120	1990
6.0	56	1510	1190	57	1550	1150
	48	1360	1450	50	1410	1400
	41	1200	1730	43	1250	1600
	38	1150	1870	39	1190	1830
	36	1120	2030	37	1160	1990

	A	B	C	D	E	F
6.5	57	1550	1190	58	1590	1150
	49	1390	1450	51	1440	1400
	42	1240	1730	43	1290	1680
	39	1190	1870	40	1230	1830
	36	1150	2030	37	1190	1990
7.0	57	1590	1190	59	1630	1150
	50	1430	1450	51	1480	1400
	42	1270	1730	44	1320	1680
	39	1220	1870	41	1260	1830
	37	1180	2030	38	1220	1990
7.5	58	1630	1190	59	1670	1150
	50	1470	1450	52	1520	1400
	43	1310	1730	45	1370	1600
	40	1260	1870	42	1300	1830
	37	1210	2030	39	1250	1990
8.0	58	1670	1190	60	1710	1150
	51	1520	1450	53	1560	1400
	44	1350	1730	45	1400	1680
	41	1290	1870	42	1330	1830
	38	1250	2030	39	1280	1890

*Increase or decrease water per cubic yard by 3 per cent for each measure or decrease of 1 m.m slump, then calculate quantities by absolute volume method. For manufactured fine aggregate, increase percentage of fine aggregate by 3 and water by 17 lb. per cubic yard of concrete. For less workable concrete, as in pavements, decrease percentage of fine aggregate by 3 and water by 8 lb. per cubic yard of concrete.

Table E-3. Suggested Trial Mixes for Air-Entrained Concrete of Medium Consistency with 3- to 4-Inch Slump.

Water-cement ratio Gal per sack	Maximum size of aggregate inches	Air Content (entrapped air) per cent	Water gal per cu yd of concrete	Cement sacks per cu yd of concrete	With fine sand fineness minimum = 2.50		
					Fine aggregate per cent of total aggregate	Fine aggregate lb per cu yd of concrete	Coarse aggregate lb per cu yd of concrete
4.5	3/8	7.5	41	9.1	50	1250	1260
	1/2	7.5	39	8.7	41	1060	1520
	3/4	6	36	8.0	35	970	1800
	1	6	34	7.8	32	900	1940
	1½	5	32	7.1	29	870	2110
5.0	3/8	7.5	41	8.2	51	1330	1260
	1/2	7.5	39	7.8	43	1140	1520
	3/4	6	36	7.2	37	1040	1800
	1	6	34	6.8	33	970	1940
	1½	5	32	6.4	31	930	2110
5.5	3/8	7.5	41	7.5	52	1390	1260
	1/2	7.5	39	7.1	44	1190	1520
	3/4	6	36	6.5	38	1090	1800
	1	6	34	6.2	34	1010	1940
	1½	5	32	5.8	32	970	2110
6.0	3/8	7.5	41	6.8	53	1430	1260
	1/2	7.5	39	6.5	45	1230	1520
	3/4	6	36	6.0	38	1120	1800
	1	6	34	5.7	35	1040	1940
	1½	5	32	5.3	32	1010	2110

6.5	3/8	7.5	41	6.3	54	1460	1260
	1/2	7.5	39	6.0	45	1260	1520
	3/4	6	36	5.5	39	1150	1800
	1	6	34	5.2	36	1080	1940
	1½	5	32	4.9	33	1040	2110
7.0	3/8	7.5	41	5.9	54	1500	1260
	1/2	7.5	39	5.6	46	1300	1520
	3/4	6	36	5.1	40	1180	1800
	1	6	34	4.9	36	1100	1940
	1½	5	32	4.6	33	1060	2110
7.5	3/8	7.5	41	5.5	55	1530	1260
	1/2	7.5	39	5.2	47	1330	1520
	3/4	6	36	4.8	40	1210	1800
	1	6	34	4.5	37	1140	1940
	1½	5	32	4.3	34	1090	2110
8.0	3/8	7.5	41	5.1	55	1560	1260
	1/2	7.5	39	4.9	47	1360	1520
	3/4	6	36	4.5	41	1240	1800
	1	6	34	4.3	37	1160	1940
	1½	5	32	4.0	34	1110	2110

*See footnote at end of table.

283

Table E-3. Suggested Trial Mixes for Air-Entrained Concrete of Medium Consistency with 3- to 4-Inch Slump (Continued from Page 283).

6.5	56	1530	1190	58	1570	1159
	48	1330	1450	50	1380	1400
	41	1220	1730	43	1270	1680
	38	1150	1870	39	1190	1830
	36	1120	2030	37	1160	1990
7.0	57	1570	1190	58	1610	1150
	49	1370	1450	56	1420	1400
	42	1250	1730	44	1300	1680
	38	1170	1870	46	1210	1830
	36	1140	2030	37	1180	1990
7.5	57	1600	1190	59	1640	1150
	49	1400	1450	51	1450	1400
	43	1280	1730	44	1330	1680
	39	1210	1870	41	1250	1830
	37	1176	2030	38	1210	1990
8.0	58	1630	1190	59	1670	1150
	50	1430	1450	51	1480	1400
	43	1310	1730	44	1360	1680
	40	1230	1870	41	1270	1830
	37	1190	2030	38	1230	1990

*Increase or decrease water per cubic yard by 3 percent for each increase or decrease of 1 n in. slump, then calculate quantites by absolute volume method. For manufactured fine aggregate, incease percentage of the aggregate by 3 and water by 17 lb per cubic yard of concrete. For less workable concrete, as in pavements decrease percentage of fine aggregate by 3 and water by 8 lb per cubic yard of concrete.

Table E-4. Approximate Mixing Water Requirements for Different Slumps and Maximum Sizes of Aggregates.

Maximum size of aggregate, in.	Air-entrained concrete				Approximate amount of entrapped air, per cent	Non-air-entrained concrete		
	Recommended average total air content, per cent	Water, gal. per cu. yd. of concrete** Slump, in.				Water, gal. per cu. yd. of concrete** Slump, in.		
		1 to 2	3 to 4	5 to 6		1 to 2	3 to 4	5 to 6
3/8	7.5	37	41	43	3.0	42	46	49
1/2	7.5	36	39	41	2.5	40	44	46
3/4	6.0	33	36	38	2.0	37	41	43
1	6.0	31	34	36	1.5	36	39	41
1½	5.0	29	32	34	1.0	33	36	38
2	5.0	27	30	32	0.5	31	34	36
3	4.0	25	28	30	0.3	29	32	34
6	3.0	22	24	26	0.2	25	28	30

Appendix F
Poured Reinforced
Concrete Construction

Figure F-1 shows what a section of wood forms would look like if you were to pour walls in an underground house. Even though the pictures in this section are of a building above ground, nothing would change. Look closely at all that support before you decide to design a house with poured walls. Figure F-2 is a view of the same wall from the inside just before the final skin of plywood was mounted in place. Figure F-3 shows the same wall at the point where the drain pipe enters the building. Figure F-4 is a close-up of what the poured concrete looks like as the skin of plywood is removed two weeks later.

Fig. F-1. Forms for a poured wall.

Fig. F-2. View of the poured wall from the inside.

Fig. F-3. The drain pipe is in place.

Fig. F-4. A close view of poured concrete.

Appendix G
How To Get Work Done

Since you are building this house yourself, you certainly don't lack energy or interest, so there are only a few things that can slow you down. These include the weather, lack of money or poor planning. The first you have no control over. Money is a personal thing between you and your bank. You know how much you have to work with and how careful you must be with it. This only leaves planning. This you have total control over. Whatever you do, don't confuse planning with designing. Designing is physically locating material items. Planning is the art of allocating your time.

ALLOCATING TIME

Design will be settled long before you actually begin to build, and you really can't do much changing after you have actually begun pouring concrete and laying block. Once your plans are approved by all necessary authorities and a building permit issued, you can consider the design phase of your project final, except in minor instances.

However, time allocation will continue from day one of construction to the time you use the last paint brush. So you see, this is the important part of building that you have control over. Time is as important as money. Sometimes it is more important because money can't buy time, but in time you can make money.

Once you're committed to building this underground house, figure on one year of continuous work on your part, maybe even more, depending on your desire to get the job done and nature's cooperation.

First of all, don't get behind before you start. Break ground in the early spring. If you wait until mid-summer, you will be in a race with mother nature to beat cold weather. Of course, I'm assuming that you're building the underground house in a region with extreme seasons, such as in Maryland. If you happen to be in the deep South or Southwest, you just have to appreciate our problems of changing seasons. The reason it is imperative to complete all concrete work and get it covered with earth is simple. Concrete expands and contracts like anything else with heat and cold. The expansion and contraction cause cracking which is a condition you need to avoid at any cost. This is why I suggest you plan carefully to be able to pour your roof slab during a time period where the temperature doesn't get below freezing or vary to extremes of hot and cold. You must realize that if your building were being built in a locale where the temperatue might reach 100°F during a day in September, but fall to 40°F at night causing a 60°F differential, cracking could develop. This condition would probably do more damage than acutally pouring concrete at 50°F and having the temperature drop to 32°F at night. The extreme differential is what causes the cracking problem and should be avoided at any cost. How's that for a planning problem?

I can't begin to tell you how to plan well personally. I don't think any book can. You are born with that ability. It's almost a talent—like singing. Fortunately, I was born with a planning ability a little above average. However, I know that if I had been even better at planning, my job would have been much easier. Just think a little bit ahead. Don't overlook little things, like small tools, nails or a water cooler.

A good example is something that happened to me. I had a gas generator on the site to use while building the scaffolding for the roof. I rushed to the site early in the morning with my power saw, started the generator and, lo and behold, I'd forgotten my extension cord. For the want of an extension cord hours of work were lost until I left the site and picked one up. Now I was ready to cut boards, and I did, for about 15 minutes. Then the generator ran out of gas. As if this wasn't bad enough, I didn't have a container to get gas in. So by the time I got a can out to the gas station, back to the site, started the generator and was ready to work, I was starved and ready for lunch. Of course, I didn't brown bag a lunch at first because I didn't realize how much time it takes to drive three miles, get a sandwich and soda and get back to work. I soon learned to brown bag a sandwich on those days in which I planned to put in a full day's work. If there is food and drink on the site, I found I could get an average of two more hours a day of working time. This really adds up in a year.

I have to admit that this day I described was unusual for me, but it is an example of what good planning and a little thought can prevent.

As for how to get additional working time, this is the eternal problem of a do-it-yourselfer. I presume the reader of this book to be working a regular 40-hour week to pay for the other 128 hours and the house. Let's also assume that you use 10 hours a week travelling to and from work and to and from the building site; you probably need seven hours of sleep a night; and time to eat. Let's say that is 10 hours a week. This gives you a total of 109 hours just to eat, sleep and work at your regular occupation. That leaves 59 hours to take care of personal business, see your family, socialize and build a house. So you see, there's really no time to waste in idle talk. You cannot waste effort of any kind.

One of the fine lines you will have to walk when building an underground house is tactfully dealing with and handling friendly, inquisitive people. This sounds like an unfriendly gesture, but it is not. Here's what I mean. Naturally an underground house is interesting. Maybe some day there will be enough underground homes around so that the majority of the public will have seen one or possibly have been inside of one. Once this happens, of course, the novelty will be missing and the curiosity seekers will not visit you, but this is definitely in the future era. Be prepared for the present day, because as the word of your innovative project spreads, your friends and neighbors will stop by to talk with you about your house. This is the problem. How do you politely keep on working while they are asking you basic questions? Remember, after a while, you have answered every question many times and explained the details to many people. But each time the questions are asked, it is the first time for your friend. Soon you will find yourself talking more than working. However, it is good to know that people are interested, so you just do your best to work and talk, talk and work. Most people will volunteer a helping hand and that is something you can always use.

INSURANCE

This brings me to another subject that should be mentioned in conjunction with an underground house. Insurance. With all the visitors you will have on your property from beginning to end, I suggest you obtain a good insurance policy covering personal liability. As I'm sure you are aware, it's awfully easy to have an injury on a construction site, especially in the early stages when the terrain is full of ditches, rubble, nails, loose timber and so on. Ask your lawyer about posting the property as a means of liability protection, but don't count on a *no trespassing sign* to protect you against a lawsuit if someone would get hurt, because it would be voided by your personal invitation to visitors or workers. You are still liable for these people. Anyway you slice it, you need a good insurance policy.

BEST WORKING HOURS

As for working time, I found the best time to work uninterrupted was to start at sunrise. I don't mean just early—I mean exactly at sun-up. Somewhere around 5 a.m. Once you get used to these early hours you'll find work progresses must faster than evening work. It's also invigorating and the sunrises seem to be good for the mind. To be honest with you, I found the weekends, holidays and vacation days to be the time to get the big jobs done. You'll find that the larger phases of construction need to be completed by continued hours of working. You can literally never get a big job done (for example, plumbing) if you do it 15 minutes at a time, one day at a time. It will take 15 minutes to get the necessary tools and supplies together and ready for work, whether that work period is to be for 10 minutes or 10 hours. This is another way to save valuable time by planning. So if you have only a few minutes to work, do something that can be completed in that time frame, if at all possible. One such job would be hanging a door. Or use the time to prepare a particular area for a bigger job. Keep this train of thought in mind and you will find you have saved days by the time you're finished building.

The early mornings and evenings I used to take care of small things. If you can arrange to take a vacation at times when a big job needs to be completed, then you are fortunate. There's nothing worse than having a complete week off from your normal job, have it pass and accomplish nothing significant. It can really demoralize and frustrate you if it happens.

Still another phase of getting work done is affected by availability of material. This is where planning far ahead on your part can prevent a problem. When ordering material, be sure to give the supplier long enough advance notice to deliver supplies by the time you need them. Don't wait until the last minute and then waste time because you are missing one 2 × 4 or one concrete block. Don't forget, you're not the only one ordering supplies. Don't expect the vender to jump at your request for immediate delivery.

If there is one way above all others to save time, it is to complete one craft throughout the house. For example, if you are putting furring strips on the block walls, start in one room and continue throughout the house. The same goes for plumbing, electric, sheet rock, painting or whatever. If you try to complete one room at a time, from scratch to final trim, and then move on to the next room, you will spend most of your time putting tools away and getting new ones out. Also by completing one job throughout the house, you become skilled at the particular job. The only thing to be said against this method is that it gets boring. For example, hammering nails for three days and then painting for three days becomes very monotonous. Jumping around can break this monotony, but it wastes valuable time.

Probably the most effective time saving idea that I will mention is following: When you have help available, whether it's a friend, relative, paid worker, young or old, male or female, don't overlook the fact that everyone can do some type of work. But, if you ask the wrong person to do the wrong job, you either have to hold their hand or do the job over. Don't ask a child to do a man's work and vice versa. This way neither person will become frustrated by being asked to do a job they can't handle. Use your labor, whoever or whenever, effectively.

About the only other thing which is really important as far as saving time and making work easier is to have work that needs to be completed inside and outside even after you are under roof. In Hartford County, Maryland, just as in most other locales, the weather varies considerable. Even in mid-winter, we have occasional time periods of warmer than usual temperature. When this happens, get something done outside that can't be completed in bad weather, because in the warm weather there are surely going to be days that rain prevents outside work. I know of people who would just cancel that day's work. But if you're really anxious to complete your house, you can always find something constructive to do whether it's winter or summer, night or day.

Appendix H
Wood Burning Stoves

Your needs and requirements must be determined before purchasing a wood burning stove. What do you expect the stove to do? How much space do you want the stove to heat? This is measured by cubic feet capacity and will probably yield only about 80 percent capacity of the manufacturing label. Too large a stove will result in inefficiency. It will never burn at the rated capacity to function best. Too small a stove will not heat the area properly and you may load it beyond its rated capacity. To determine the room capacity to be heated, multiply the length by the width by the height.

How long do you want the fire to burn? Stoves are traditionally built to hold a fire only a couple of hours for every one loading. An airtight high efficiency stove will hold a fire for eight to 12 hours. They need cleaning more often and cost more. This type stove is the least expensive on a long term basis.

How important is appearance? Some efficiency stoves are not very attractive and some inefficient stoves are extremely attractive. There are a few that will give you both.

If you are planning to burn coal in addition to wood, be sure the stove can handle the heat. Some wood stoves need only a coal grate to burn coal.

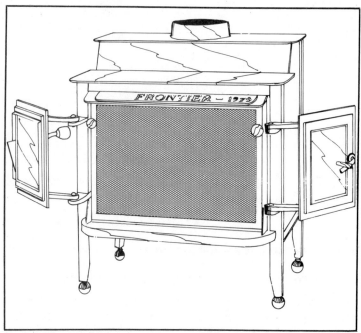

Fig. H-1. Wood stoves help to save money and energy.

Power loss may deprive you of cooking with electricity. A cook top stove can be pretty handy under those circumstances. Some units have a usable top to be used for cooking.

Stoves with a visible fire are equipped with screen or glass front. Such a stove is diminished in efficiency. If you want a stove with a visible fire, be prepared for the loss in heat.

Some stoves are equipped with a low voltage electric blower to circulate the heat. If this is your choice, be sure to get one with a ventilated motor. They will last longer than the enclosed type.

Stovepipe is important. The stove manufacturer will recommend the size to be used. The heavier the metal, the longer it will last. Twenty-four gauge is standard. The lower the number of gauge, the thicker the metal. Be sure to install a non-combustible floor under the stove. Cover at least 6" beyond the dimensions of the stove and at least 1' on the fuel feeding side. The manufacturer will recommend the minimum distance the stove should be from the wall.

Wood is becoming one of our favorite alternative fuels. Attracted by the prospect of saving energy and money, thousands of homeowners all across the nation are installing wood stoves. Unfortunately, many of these stoves are being improperly installed or carelessly operated so wood stove fires are becoming more and more common. Wood stoves are not inherently dangerous. Most of these fires could have been prevented by common safety measures.

STOVE MODELS

There are about 1,500 new wood stove models to choose from. Some are unattractive and some are beautiful. Some reflect the late 19th century styling (Figs. H-1 through H-12).

Fig. H-2. You can cook on this wood stove.

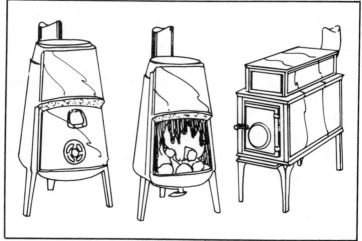

Fig. H-3. Wood stoves come in numerous styles.

A thorough inspection is necessary to restore an old stove back into service. If the firebox and grates are in good condition, the stove is as good as new. Although cracks can be repaired, it may be a danger because the heat inside the firebox is so intense that a bad patch presents a danger, possibly causing the house to burn down.

The Franklin Fireplace. These units have doors which will permit some degree of control, and can raise the efficiency level to about 15 percent to 20 percent.

Pot Belly-Parlor Stove and Box Heaters. These are closed stove type heaters with manual damper control and can be as much as 40 to 50 percent efficient with proper damper setting.

Air Tight. These are thermostatically controlled units with 50 to 60 percent efficiency. Any desired temperature level can be maintained because of precise combustion control and draft intake. One fueling can last as long as 14 hours.

CAST IRON VERSUS STEEL

Either cast iron or steel is accepted as stove building material. Cast iron's reputation of holding heat longer is a myth. It stems from the days when cast iron stove walls were thicker than steel stove walls. Given equal thickness, either metal will give identical performance. Light gauge metal without fire brick lining can be very dangerous in any unit and will have a limited life. The heating capacity of a stove is governed by the weight, mass or thickness, not the metal type.

A used stove should be checked for cracks and repaired with stove cement. If the cracks are wide open, they should be welded. Buy a substantial good stove with good workmanship and design. The thicker the metal, the longer the stove will last. The high efficiency stoves are more expen-

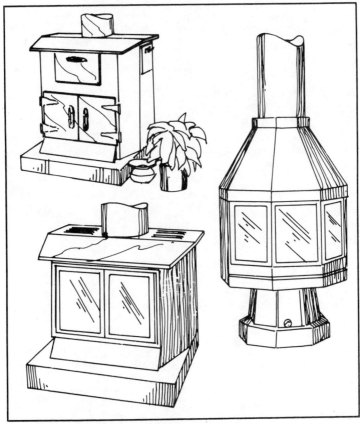

Fig. H-4. Avoid fires by installing wood stoves properly.

Fig. H-5. This stove serves as a fireplace (courtesy of Martin Industries).

Fig. H-6. Get a wood stove which goes well with the room's decor (courtesy of Martin Industries).

Fig. H-7. Thoroughly inspect your stove before installing it (courtesy of Martin Industries).

Fig. H-8. An older wood stove design (courtesy of Martin Industries).

sive to buy but are less expensive to operate because they deliver more heat.

A closed wood burning stove has a different effect on heat. The stove is completely enclosed except for air entering the stove for control. Heat from the stove will heat the room with little or no loss in addition to reducing the amount of wood burned.

A fireplace stove might be the happy medium, which is nothing more than a wood burning stove inserted into the firebox using the flue as a chimney. A fan built within the system will help circulate the heat from the stove into the room. The surface temperature is about 400°F. This thin layer of hot air around the stove is called *Langmuir's film*. The fan will blow Langmuir's film into the room away from the stove. Any heat not able to radiate away from the stove will be lost through the chimney.

Adequate wall, floor and ceiling clearances must be provided for safety. Place the stove at least 36″ from a combustible or unprotected wall. Sheet asbestos with a 1″ clearance from the wall will allow 18″ clearance. If 28 gauge sheet metal is used in lieu of asbestos, the clearance can be 12″.

Floor clearance is generally not as severe a problem because the ashes reduce radiation and act as an insulation. However stoves with 18″ legs or more should sit on 28 gauge metal. Shorter legs will require a fireproof floor such as concrete, brick, tile or asbestos. Stovepipe installation

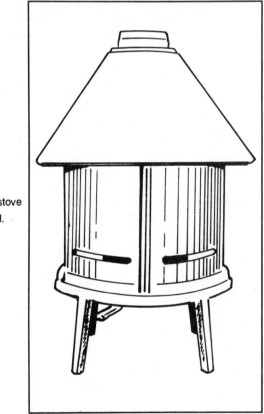

Fig. H-9. This wood stove is uniquely designed.

Fig. H-10. A closed stove type heater.

should have a minimum of turns or bends. Plan to replace the pipe about every two years.

The connector must have an internal cross-sectional area of not less than that of the flue collar of the stove. The connector must be straight and as short as possible, preferably with not more than one or at most two sweeping 90° elbows. If you connect into an existing chimney, be sure the flue lining is clean and not cracked. The local fire department or building inspector will inspect the chimney. It should be inspected every year.

Keep firewood, kindling and newspaper well away from the stove or fireplace. The sides of the stove should be at least 36″ away from any combustible material. Use only the fuel which the stove was designed to burn. If a wood burning stove does not burn coal, and if a coal burning stove does not burn wood, do not burn both unless you purchase one that will burn either. Keep a fire extinguisher handy. Install smoke and fire detectors and finally plan an escape route from your house.

SMOKEPIPE INSTALLATION

The smokepipe is connected to a thimble or flue ring built into the wall extending to the flue lining. The rings are available with inside diameters of 6″, 7″, 8″, 10″ and 12″, and lengths of 4½″, 6″, 9″ and 12″. When installed into the wall, these rings should be airtight with a full bed of mortar around

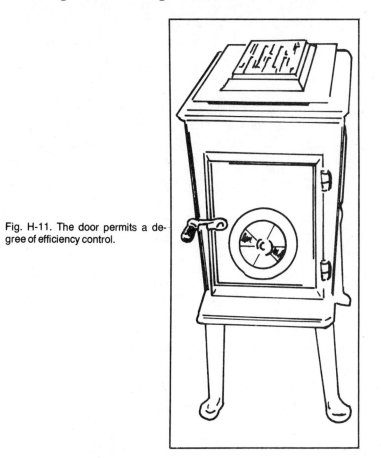

Fig. H-11. The door permits a degree of efficiency control.

Fig. H-12. A large Franklin fireplace unit.

the ring. When installing a metal stovepipe, it should be installed no farther than the inner face of the flue lining with an airtight joint, along with boiler putty or asbestos cement. The top of the smokepipe should not be less than 18″ below the ceiling and no wood should be placed within 6″ of the thimble.

An improperly installed smokepipe from a wood burning stove is a dangerous fire hazard (Fig. H-3). Do not extend the pipe too far into the flue. A metal or clay collar should be built into the wall or brickwork to receive the smokepipe and a collar slipped over the pipe. Smokepipes should be kept at least 10″ away from the woodwork of the house.

Some building codes will require fireproof ceilings and walls in stoved rooms. Some communities may require local building permits for installation. The building inspector will offer assistance for installation if requested.

The clearance of chimney connector should be not less than 18″ from the nearest combustible material. Where connected to a masonry chimney, a chimney connector is

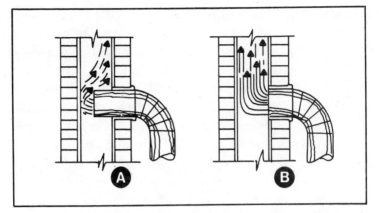

Fig. H-13. (A) Improper stovepipe connection. (B) Proper stovepipe connection.

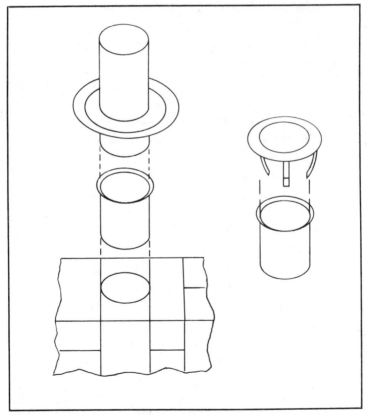

Fig. H-14. The thimble is removed and a chimney connector is installed.

required to extend through the masonry wall to the inner face of the liner but not beyond the inner surface. The connector must be mortared to the masonry wall. In lieu of this, a thimble may be connected to the masonry wall of the chimney without mortar (Fig. H-14). If mortar is used, it must be the high temperature type. If the thimble is used, the joint must be tight to prevent back draft smoke from leaking through. A damper should be installed on the stove pipe.

A manual operated damper must be installed close to the stove. It may be horizontal or vertical. An alternative is an automatic draft regulator.

SETTING A STOVE

A wood stove must be a certain distance away from combustible materials to prevent fires from radiated heat. If your stove is listed by a national laboratory such as Underwriters' Laboratories, Inc., follow the clearances they specify. If your stove is not listed, diligently adhere to the specifications given here. Be aware that the sides, bottom and stovepipe attain quite different temperatures while the stove is in operation, so there are quite different specifications for each of them.

The sides of an unlisted stove should be at least 36" from any combustible materials. A foot or two of clearance will not do. Even though very high temperatures are needed to ignite most combustible materials, over time, high temperatures can change the composition of the material by slowly darkening it so that it accepts more and more radiated heat. Finally, it may begin to smolder and burn at temperatures as low as 200 to 250°F. These temperatures are easily reached by an unprotected wall exposed to a stove without adequate clearance, so strict adherance to the clearance standards are advised.

An unlisted stove may be positioned closer to a wall than 36" only if a noncombustible material is spaced at least 1" away from the wall to allow air to circulate behind the

Type of protection	Stove Sides	Stovepipe
Unprotected	36″	18″
1/4″ asbestos millboard	36″	18″
1/4″ asbestos millboard spaced out 1″	18″	12″
28 ga. sheet metal on 1/4″ asbestos millboard	18″	12″
28 ga. sheet metal spaced out 1″	12″	9″
28 ga. sheet metal on 1/8″ asbestos millboard spaced out 1″	12″	9″

material and carry heat away. But placing a noncombustible material directly on the wall has no real protective value. The material will easily conduct heat to the wall behind it creating dangerous conditions.

Asbestos millboard spaced 1″ away from a wall will allow an 18″ stovewall clearance. If 28 gauge sheet metal is installed in the same manner, the acceptable clearance is only 12″.

Remember that these clearances are for all combustible materials. This eliminates some dangerous but common practices such as stacking wood or paper next to a stove or moving a sofa temporarily closer. It may be difficult to install a stove where it is convenient, aesthetically pleasing and safe. Don't compromise. Your priority is safety.

At one time, separate standards were specified for stoves with sheet metal cabinets (also called circulating stoves), but the National Fire Protection Association now recommends that all wood stoves be treated the same (Table H-1 and Fig. H-15).

FLOOR PROTECTION

Safe floor clearances are substantially less than those for walls because the heat radiated from the bottom of a stove is generally less than from either the sides on the top. During a fire, ashes fall to the bottom of a stove. This has an

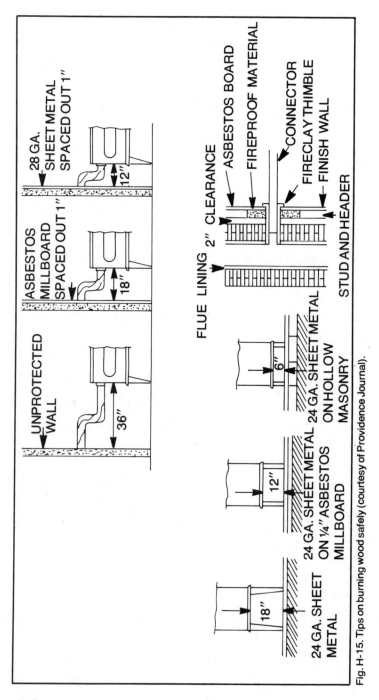

Fig. H-15. Tips on burning wood safely (courtesy of Providence Journal).

Table H-2. Leg Length and Floor Protection Information.

Leg Length	Protection Needed
18" or more	24 ga. layer of sheet metal
6" to 18"	24 ga. layer of sheet metal over 1/4" layer of asbestos millboard
6" or less	Use 4" of hollow masonry laid to provide air circulation through the masonry layer covered by a sheet of 24 ga. sheet metal.

insulating effect, resisting the flow of heat downward. This process may be started in a new stove by placing a layer of sand in the bottom.

Clearances for proper floor protection are classified according to leg length, and are listed in Table H-2. Floor protection materials can be covered with attractive *materials* such as stone, brick or tile as long as they are noncombustible. Do not use these materials instead of an asbestos sheet or metal plate.

Falling embers and sparks present an additional safety problem that is often ignored. Avoid this potential problem by extending the floor protection 18" from the front of the stove and 6" around the sides and back. This affords a reasonable amount of protection, but you should still take care when loading and tending the stove. Make sure that ashes and hot coals fall only on the protected area.

STOVEPIPE

Stovepipe and chimneys, although they are often mistakenly used interchangeably, are two completely different things. When a stovepipe is used for a chimney, dangerous conditions may occur. Creosote may build up rapidly and the wind and rain may soon corrode the pipe. Consequently, if a stovepipe fire should occur, there will be little to contain it. Chimneys, in contrast, keep the smoke and gases hotter to prevent the buildup of creosote. They are made of corrosion resistant materials so they do not need frequent replacement, and most chimneys are able to contain a fire.

A stovepipe must be installed so that there is a good draft to carry the hot gases away quickly and safely. This can be a tricky proposition unless you follow these guidelines carefully:

• Keep the stovepipe short.

• Keep turns and bends to a minimum.

• The horizontal portion of the stovepipe should be no more than 75 percent of the vertical portion.

• The stovepipe should enter the chimney well above the stove outlet. Horizontal portions of the stovepipe should rise at least ¼" per foot.

Stovepipes come in different diameters and gauges. Measure the pipe connector on the stove and get the proper size stovepipe. Pipe gauge 24 or thicker is a good investment because it will last longer, lower gauge numbers indicate thicker metal. The stovepipe pieces should be fitted together tightly, then permanently joined with two or three sheet metal screws to prevent the stovepipe from being shaken apart during a chimney fire. Since stovepipe cannot be expected to last long, inspect it regularly and plan on replacing it every two or three years.

All stoves should have some sort of damper. Dampers are used to control the air flow to a stove, and in the event of a chimney fire, they should have the capability to shut off the air flow to the chimney completely. Generally, airtight stoves have sufficient dampers built in, but any non-airtight stove should have a damper in the stovepipe. It is a good idea to install a second damper in the stovepipe as a safety control. Most of the dampers sold in hardware stores are not solid. Although they can be used to control air flow they cannot shut it off completely. You may have to replace this damper with a solid piece of metal cut the same size.

Care must be taken when attaching the stovepipe to a chimney, and to allow for proper cleaning of the chimney.

Stovepipe clearances must be maintained from all combustible surfaces and, in a masonry chimney, a special connector thimble must be installed. The stovepipe should fit into this thimble, and the thimble must fit into the chimney and be permanently cemented in place. Do not extend the thimble into the flue.

STOVE MANUFACTURING MATERIAL

There are three choices: cast iron, plate steel and sheet metal. Cast iron stoves are made by pouring molten pig iron into a mold forming the component parts of a stove. There pieces are machined to fit, sand blasted and acid dripped (pickled) to remove scale and corrosion, assembled and either enameled or blackened.

Plate steel stoves are heavy sheets of steel (typically ¼"-½" thick) cut to size and then welded together. They are wire brushed or pickled and then blackened. Sheet metal is thin steel sheet (less than ⅛" thick) that is either welded or bolted to form the stove. The material difference between plate steel and cast iron is different enough not to warrant significant concern. The thicker the metal, the longer the stove will last. Both materials retain roughly the same amount of heat per pound and both stove types cost roughly the same. One difference is that a cast iron stove can be dressed up with designs in the casting.

A well made stove will have clean castings, smooth welds and good workmanship. Some stoves have firebrick or metal plates to prevent burnout; this may increase the lifetime of the stove and also increases the *thermal mass* (the storage medium for the heat- a 500-pound stove can continue to give off usable heat four hours after the fire is out).

In general, the material a stove is made from is secondary to the workmanship and design. Sheet metal stoves are made from sheet metal (thin sheets) that burn out fast. They may be inexpensive, but they will need frequent replacement.

Plans are available for constructing a stove from a discarded oil drum. They are ideally suited for heating a garage, barn or camp, but are not recommended for extensive home heating. The sides get red hot.

STOVE DESIGNS

The design of a stove in any given category may vary significantly from stove to stove, yet the basic principle remains that of extracting heat from the fire. Low efficiency stoves (20-30 percent efficient) are technically the simplest. They have a straight air path either across or through the fire allowing sufficient air for combustion of the wood (primary air) with no accommodations for additional air to support the flame combustion (secondary air). They have excessive air leaks due to unsealed beams resulting in poor fire control. Simple box stoves, Franklin stoves, chunk stoves, pot belly stoves, parlor stoves and sheet metal stoves are representative of the category.

Medium efficiency stoves (35-50 percent efficient) are more sophisticated. They have better control of the amount of primary and secondary air and are built without excessive air leaks. The common term for this category is *airtights*. Air flow to the fire is controlled to ensure maximum utilization of the burn. A thermostat is commonly utilized to insure a constant burning rate.

High efficiency stoves (50-60 percent efficient) are the most technically sophisticated. They regulate air flow as in the simpler airtights but employ baffles, long smoke paths and heat exchangers to extract as much heat as possible from the fire. These stoves are the most expensive to buy, but they deliver the most heat per unit of fuel.

When selecting a stove, the following considerations are rated in priority order:
• Decide on what your end use will be.
• Decide on location(s) of the stove(s).
• Select the correct size for your application.

- Pick a general type that best fits your use.
- Take to a stove dealer, tell him what you want to buy and why.
- Pick a stove that fits your use and taste, not just your pocketbook. Consider the purchase of a stove a major investment that will pay a handsome dividend over its lifetime.

Appendix I
Wood As A Fuel

The phrase "please turn up the heat" requires more than just the flick of a switch. It means going outside to get wood to restock the woodbox, removing the right amount of ashes, mixing the right seasoned wood for a long even burn, adjusting the air flow and feeding the fire for an even burning lasting heat. No chemicals should be used in a wood stove such as commercially produced sawdust logs. No garbage should be used for kindling. Wood storage requires a great deal of space. It tends to be somewhat messy with dirt, wood wastechips, sawdust and insects that are part of burning wood.

Household solid fuel combustion is a complex chemical reaction between oxygen and fuel. The result is heat and gases. Several conditions must be followed to assure maximum combustion:

• Continuous supply of fresh air.

• Proper amount of fuel in the combustion chamber not enough to overheat, but enough to avoid a cool flue liner which will result in poor draft and buildup of residues and creosote.

• Proper fuel burning to avoid waste and toxic products of combustion, such as carbon monoxide.

• Removing unwanted products of combustion.

Combustion occurs when all the fuel is oxidized by the air. Stoves and fireplaces are designed to operate with a considerable amount of excess air to prevent intermittent periods of incomplete combustion. When incomplete combustion occurs, carbon monoxide is produced.

WOOD BUYING

Wood buying is not a simple matter. In order for you not to "get burned," decide on an amount of wood, know the moisture content, the degree of preparation and the price. Wood is sold in units of cords (Fig. I-1). A full cord measures 8' long, 4' high and 4' wide. A face cord has wood lengths of 12" to 16" to 24", which makes it less than a full cord and the price should be less; 24" long logs is a full cord, 16" long logs is one-third less a full cord and 12" long logs is ¼ less a full cord. When purchasing firewood, be sure the stacking of the logs is fairly tight so you will not be shortchanged. A straight cord is made up entirely of one grade or species. A mixed cord is made up of several different wood species.

The kind of wood is important. The best firewoods are the hardwoods because they produce more heat when burned. As a rule, hardwood trees are those that shed leaves and change seasonally. Softwoods generally are evergreens with needles.

A full cord of wood can weigh as much as 1 ton. From 3 to 10 cords of wood may be required to heat a house during one winter season. Beware of buying a truck load of wood. A standard pickup truck will hold only about one-third or one-half a full cord. The average wood burned per year is about 1 to 2 cords for fireplaces, 4 to 6 cords for stoves and 6 to 8 for woodburning furnaces. Wood heating value is the most important characteristic we should be concerned about. The more dense the wood, the more heat it will give off when burned. Woods are broken into hardwood and softwood with heating values between. Following are the groups broken down:

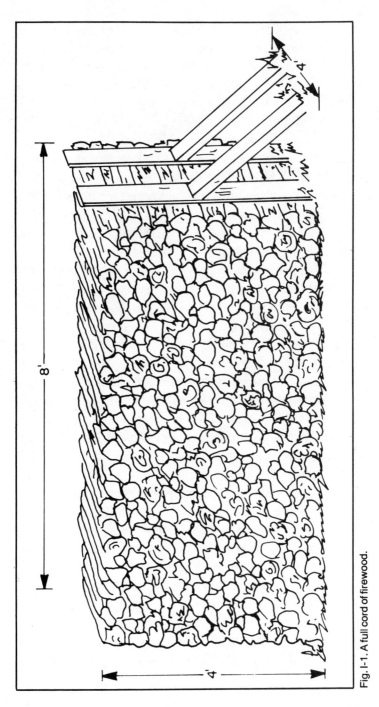

Fig. I-1. A full cord of firewood.

• The woods with the highest heat values, 28-30 million BTU per cord, are apple, birch, black locust, hickory, hop horn bean and white oak. One cord of these woods is equal to approximately 191 gallons of fuel, 228,000 cubic feet of natural gas and 5020 KWH of electricity.

• 25-26 million BTU per cord include ash, beach, red oak, sugar maple and yellow birch. These equal approximately 176 gallons of fuel oil. 211,000 cubic feet of natural gas and 4,600 KWH of electricity.

• With 21-23 million BTU per cord are black cherry, elm, grey birch, red maple, tamarack and white birch. These equal about 150 gallons of oil, 180,000 cubic feet of gas and 3971 KWH of electricity.

• The lowest woods in heat value are alder, balsam, fir, basswood, adar, hemlock, poplar, spruce and white pine, equal to about 108 gallons of oil, 129,000 cubic feet of gas and 2800 KWH of electricity.

Be sure the firewood used, or cut, is dry before burning. (Fig. I-2). Moisture content should be about 20 percent and it takes fresh wood about eight months to reach that level. Only then will you achieve the maximum heat from burning wood. If you are still not sure of the dryness, take two pieces of wood and knock them together. Dry wood will make a clear almost hollow sound. Wet wood will come together with a solid firm thud.

The BTU value of woods vary according to the density. One-fifth of the wood is wasted by moisture evaporation changing into steam. Half the weight of freshly cut wood may be water and so-called dry wood may still contain about 20 percent moisture. Water must be driven out of wood before it will burn. The greener the wood, the more heat loss up the chimney. The temperature of the flue normally runs about 400° and any water will turn into steam. This steam condenses on the inside wall of a cold chimney and deposits creosote on the walls. If you must burn green wood, burn it with a hot fire.

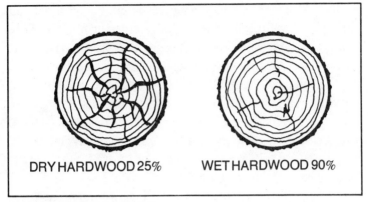

Fig. I-2. Moisture content of wood.

WOOD BURNING

Wood will burn in several stages (Fig. I-3). Heat is first applied to the wood by burning kindling and paper. This raises the temperature of the wood and drives out the moisture. At this stage, the wood is absorbing heat and raises the wood temperature until it reaches distillation. The fibers of the wood give off gases which burst into flame. This is known as "flash point," which occurs when the temperature of the wood reaches about 480°F. After the gases are burned, charcoal remains which when burned gives off a high heat output. As the charcoal continues to burn, carbon is consumed leaving ashes which will not burn. When wood is burned, it releases carbon dioxide, water and ashes. If the burning process is complete and efficient, the energy is released as heat and the original components are recycled back to the earth. There is no pollution from harmful chemicals (Figs. I-4 and I-5).

Wood burns with a higher flame because of the high percentage of volatile matter. The combustion space above the burning wood must be greater than needed for coal. Most of the air mixture is introduced around the burning wood to assure complete combustion and prevent unburned gases from entering the chimney. If a moderate sized hot fire is maintained, any creosote or gases that enter the flue as a result of incomplete combustion will be exhausted out through the hot chimney. A cooler smoldering fire will cause these gases to condense and accumulate on the wall of the flue lining, causing a possible chimney fire. A shallow bed of ashes should be left in the stove or fireplace to provide a heat reflecting surface.

Use the same principle for starting a coal fire as a wood fire. Once a draft has been created in the chimney, more coal may be added to the fire. The damper or draft must be fully open when starting a fire.

Wood should be stored in a dry location before use, and should be stacked so that both ends of the logs are exposed to

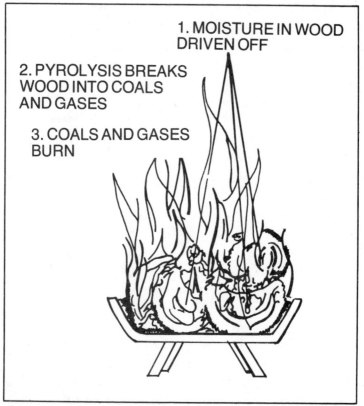

1. MOISTURE IN WOOD DRIVEN OFF

2. PYROLYSIS BREAKS WOOD INTO COALS AND GASES

3. COALS AND GASES BURN

Fig. I-3. Three stages of combustion. Three essentials are required to begin the combustion process. You need fuel, air and a high temperature to start and keep the fire going.

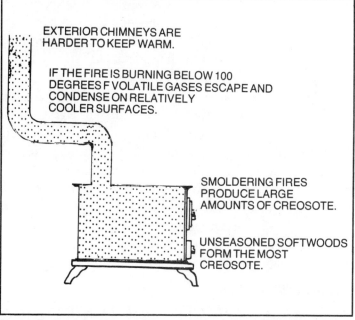

EXTERIOR CHIMNEYS ARE HARDER TO KEEP WARM.

IF THE FIRE IS BURNING BELOW 100 DEGREES F VOLATILE GASES ESCAPE AND CONDENSE ON RELATIVELY COOLER SURFACES.

SMOLDERING FIRES PRODUCE LARGE AMOUNTS OF CREOSOTE.

UNSEASONED SOFTWOODS FORM THE MOST CREOSOTE.

Fig. I-4. Burn wood properly. If the fire is burning below 1100°F, volatile gases escape and condense on relatively cooler surfaces. Smothering fires produce large amounts of creosote. Unseasoned softwoods form the most creosote.

the air. Do not burn waste paper like gift wrapping in the fireplace.

Firewood would be dried before burning. This involves loosely stacked rows to allow air to circulate for 4 to 12 months. If the logs are split, less time is required for drying. A close look at the ends of the log will determine the extent of drying. A well seasoned log will show cracks or splits running at right angles to the rings. This is caused by shrinkage as the moisture leaves the wood. If you attempt to burn green or unseasoned logs it becomes difficult to ignite the logs. If successful in igniting, it will not produce as much heat as seasoned wood. It also builds up creosote in the flue lining which is one of the causes of chimney fires. Even with dry seasoned firewood, creosote can build up from a continuous slow burning overnight fire. A frequent hot fire will retard creosote buildup.

Stoves and fireplaces can be altered to burn coal or wood. A stove and chimney connection used to burn coal should be of heavy metal thickness. Either a stove or fireplace requires grates in burning coal. Wood burning usually requires andirons.

Open fires like the sun give off radiant heat, which is the most comfortable of all types of heat. A draft free grated area requires low temperatures of radiant heat for comfort. Oxygen must be provided to keep the fire burning. Normal air leakage through the house (infiltration) is usually sufficient. If the house is extremely tight with insulation, a window may need to be opened slightly to supply the needed oxygen. Do not burn wood covered with shellac, paint or other preservatives. These are highly flammable and can produce chimney fires or toxic gases.

ADVANTAGES OF WOOD

If the price of fossil fuel is making you consider wood, you are not alone. Over 50 percent of cold climate homes are already using wood either to supplement or totally supply heating needs. Historically, wood has supplied up to 90 percent of our energy needs, but as fossil fuels became cheaper our reliance on wood as a fuel declined to less than 1 percent by 1970. After the 1973 oil embargo and the increasingly high cost of "clean-modern fuels" began ruining budgets, more and more people began returning to wood as a fuel in 1976. Over that time span, the stove business suddenly became a multi-million dollar industry. In 1973 there were less than a dozen stove manufacturers left in America. Now they cannot be counted.

Wood is a form of solar energy. Sunlight, through photosynthesis, turns carbon dioxide and water into organic material. When the tree dies, bacteria converts it back to carbon dioxide and water with the nutrients returning to the soil. When a tree is burned, it releases the same ingredients: carbon dioxide, water and ashes. If the burning process is complete and efficient the energy stored is released as heat

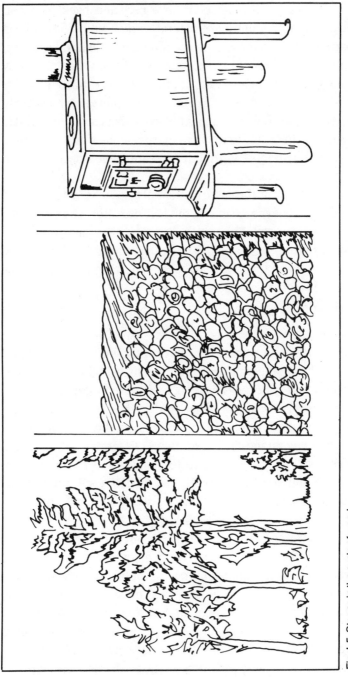

Fig. I-5. Stages in the cycle of wood.

and the original components are recycled back to earth. There is no disruption of natural cycles, no pollution from harmful chemicals, just the speeding up of a natural process. Trees are renewable. As poor quality and commercially undesirable trees are removed from the forest for firewood, new growth occurs. As long as the forest is properly managed, and only the amount of wood that can be replaced is taken, little or no significant environmental damage occurs. By encouraging new growth a greater diversity of wildlife habitats are created.

Because of the environmentally soft characteristics of wood, it has become one of the favorite alternative fuels for heat. It cannot totally replace oil, coal or nuclear power in the energy future of heating, but it can complement the energy mix and make the land less dependent on imported, expensive and increasingly scarce fuels.

WOOD PREPARATION

Preparing wood is labor intensive and the price is largely dependent upon how much work you are willing to do yourself. If you want to have your firewood split into fine pieces and stacked near your doorstep, expect to pay premium prices. If you want to go the cheapest route, cut the wood yourself from standing trees. Wood can be bought in almost any stage between these extremes. It is fairly common to buy wood in 4 to 16′ lengths and do some of the cutting and splitting yourself. Wood delivered in this form usually costs about half that of wood cut and split, but involves about eight hours of hard work.

There are no real "deals" in buying wood, just your sweat for the dealer's sweat. If you enjoy cutting and splitting wood as many people do, you can be paid handsomely for your recreation. Other sources of wood may include scavenging the local tree dump, picking broken pallets at a local mill or warehouse, purchasing ends, slabs and other waste from a local sawmill, buying cutting rights from a private landowner (many times done as shares—one cord for

**Table I-1. Seasonal Moisture Effects,
with Relative Heat Value and Moisture Content.**

Time	Relative Heat Value	Moisture Content
Green in fall, winter or spring	89	80
Green in summer	93	65
Trees leaf-felled in summer after 2 weeks	96	45
Spring wood seasoned three months	97	35
Spring wood seasoned six months	98	30
Dry wood seasoned twelve months	100	25

you and one for him), and cutting wood on state and federal lands. The possibilities are endless. The only determinant is one's ambition and resourcefulness. See Tables I-1 and I-2.

The same type of wood will vary from place to place on price, all according to labor costs, transportation costs, and supply and demand. Shop around, try to buy when demand is low and be sure you understand exactly what you are buying.

Table I-2. Labor Involved in Various Methods of Buying Firewood.

Method of Sale	Labor Involved
Truck load of green logs	Eight hours/cord
Truck load four foot rough split wood	Six to eight hours/cord
Cut and Split, you pick up	One to two hours/cord
Cut split and delivered	0
Stumpage	Eight to twelve hours/cord
Slabs, ends, etc.	Varies
Pallets & scrap wood	Varies

OPERATING A STOVE

A wood stove is not like a conventional oil or gas heating system. It can't be forgotten. There is no electronic wizardry to do the thinking for you. It needs periodic attention. Still most people consider this part of the lure of heating with wood and they thoroughly enjoy the stoking, poking and tending that their stove demands. Operating a stove is not as simple as setting a thermostat, but it is a subtle blend of science and art. The science is the thing we learn from experience and the art comes only from time and practice.

Combustion is the process that rapidly turns the chemical energy stored in wood into light and heat. This process is actually quite complicated but it can be broken down into three separate stages. First, the moisture in the wood must be driven off. Then, at high temperatures, a reaction called *pyrolysis* occurs which breaks the wood into coals and gases. Finally, at still higher temperatures, the coals and gases burn. Complicating the picture is the fact that all of these steps may be occurring at the same time in different stages and parts of the fire. Generally, a fire must go through all of these stages at one time or another.

Obviously you need a match to start the fire and high temperatures to keep it going, but you also need fuel and air. Learning to operate a stove is simply learning to control these three variables: temperature, oxygen and fuel.

The air controls not only determine the rate of burning. They also determine the efficiency of the combustion process and affect the third essential for combustion, temperature.

Turning the damper down, as is often done for long burns, results in a cool fire. Because of the low temperatures, the gases produced by pyrolysis leave the fire unburned, and a potentially valuable amount or heat is lost. Contrast this with a fire burning with the damper wide open. The fire is hot enough to burn the gases, but the temperatures create a draft so strong that heat is lost up the chimney before it can be transferred to your home. It seems like there is no way to win, so a compromise is necessary. All we can tell is that the best position is somewhere in between these two. A little experimentation will help to discover that position best for your particular stove.

FACTORS AFFECTING WOOD FUEL QUALITY

A great deal of attention is usually paid to purchasing wood and rightfully so. You should be aware of the wood species, which determines the energy content, and how long it was seasoned, which gives a good indication of the moisture content.

Hardwoods are generally preferred over softwood because they pack more energy into the same volume as softwoods. Actually all wood has the same amount of energy per pound, but hardwoods are more dense so they have more energy in a given volume. For example a cubic foot of poplar weighs 26 pounds compared to 39 pounds for a cubic foot of sugar maple. Sugar maple has 1.5 times the energy of an equivalent volume of poplar. Since different species have different amounts of energy, you can match the amount of heat you need to a wood species. In reality, we need not be quite that precise. On cold days burn hardwoods, on warm days burn softwoods, and on moderate days a mix of hardwoods and softwoods will do. In time you will learn the different species of wood, their heating qualities, and be able to match them to your needs.

The moisture in the wood is a second important factor affecting fuel quality. Burning wet wood wastes energy because every bit of moisture in wood must be evaporated before it will burn. Evaporating this moisture takes energy. Naturally, the wetter the wood the more water to be evaporated and the more energy you lose. Burn wood that has dried for at least six months and your fire will start more easily and give off more heat.

The second essential for combustion, oxygen, is controlled by the damper or air inlet of the stove. There are many different types of dampers and air inlets, but all of these have the same function. They allow you to control the flow of air going to the fire. Shut the air controls way down and the fire will burn slowly. Open them and the fire burns quickly.

STARTING A FIRE

Nothing can be quite so frustrating to a beginner as failing to start a fire. Once you learn the basic principles, nothing could be easier. Again the three essentials, fuel, oxygen and temperature, come into play. To start a fire you need dry fuel, plenty of oxygen and high temperatures.

—Open the dampers and air inlets on the stove all the way.

—Crumble or shred newspapers and place them in the stove. Burning these will provide temperatures needed to ignite the wood.

—Stack small pieces (up to 1″ in diameter) of very dry kindling on the paper, log cabin or teepee style, to insure that plenty of oxygen will create circulation in the fire. Softwoods like pine and cedar split into thin pieces make the best kindling. Any wood will do if dry and finely split. If you use your stove intermittently, it will be wise to secure a supply of pine scraps as kindling.

—Light the paper, wait until the fire is blazing and then add larger pieces of fuel. Stack them so that air can circulate freely around them.

—After 10 minutes or so, close down the air controls to the desired location.

MAKING FIRE LAST

Most people seem to enjoy tinkering with stove during the day or evening but when night comes, they get down to business. No one likes to wake up to a cold house. Learning to make a fire last is essential.

Set your heating system thermostat at 55° to 60°F. so that the heating system will take over when the fire goes out. Overheat your home to perhaps 80 to 85°F. Even if the house cools 15 or 20° by morning, the temperature will still be a comfortable 60°.

Glossy magazines contain a lot of filler which does not burn and leaves behind a large amount of ashes. Lay 15 or 20 sheets of such paper on the fire. It will burn, leaving a thick layer of ash on top of the wood. This reduces the amount of air that can get to the fuel and slows down the burning rate. When splitting wood, you will come across large knots that defy hammer and wedge. So be it. These will insure they burn slow. Save these and your densest woods for the coldest nights.

Close the dampers for long burns. Some of the heat loss due to unburned gases will be compensated for by better heat transfer between the fire and the stove walls. Remember that these unburned gases end up as creosote in the stove pipe or chimney. Periodically check your flue or pipe and clean as necessary. Try to begin a new burning cycle at the beginnings of long periods of inattention. Before bedtime all that should remain of the previous fire is a bed of coals. This insures that a fuel change will last for the maximum period of time. If seasoned wood is scarce, mixing green wood with dry wood will extend the wood pile and make the fire last a little longer. This technique should be approached with caution as green wood produces excessive creosote, and even more so when the stove is tightly dampered. If forced to use this method, it is good advice to start a good hot fire every morning to burn off the creosote formed the night before. Obviously the alternative to making a fire last is to get up early in the morning to stoke the stove. Then go back to sleep.

One of the by-products of burning wood is ash. Eventually you will have to remove some or it will fill you stove. Remove ash with a shovel and place it in a heavy bucket.

Assume that there are still hot coals mixed in with the ashes (there usually are) and do not place them in the garbage in or near anything combustible. You may wish to save the ashes in a metal drum. They make excellent fertilizer and can be used to give traction on icy walks and steps. They can even be used to make soap.

WOOD STORAGE

Split as much wood as possible. Remove the bark because it seals the wood and retains the moisture, preventing the logs from drying out. Small diameter pieces need not be split because they dry quickly. Whether you cut your own wood or buy it, try to get as many split logs as possible in your wood pile.

Stack the wood neatly, bark side up, in an area with plenty of sun, the logs in length parallel to the prevailing winds. This will help to dry out the logs. Keep the bottom of the pile off the ground by stacking the wood on rot resistant lumber such as 2 × 3″. A sheet of plastic on the ground under the log pile will help to keep ground moisture from penetrating the logs. Cover the log pile with a protective cover such as tar paper or plastic film to guard against rain, snow and ice.

Check to see if insects or spiders are beginning to nest in the wood pile. If this happens don't use insect repellents or pesticides on the wood. When the logs are burned, this will cause dangerous fumes. Move the pile to another location where it is drier and there are stronger winds. Keep the area clean to discourage termites and carpenter ants or other wood eating pests. If you do see them, take precautions so they will not find their way into the house or other buildings. Never place a log on the ground in contact with the house or building that might provide a home for the insects. After about three months of temperatures around 50°F., the wood will be as dry as this method will allow it to be. Store the wood where it will protected against the weather. Don't store wood inside the house. Carry in as many logs as the fire will support. Any dormant insects nesting in the logs will come alive from the heat inside the house.

Appendix J
Sample Of A Building Code

ARTICLE 10: CHIMNEYS, FLUES AND VENT PIPES
Section 1000.0 General

1000.1 Scope. The provisions of this article shall control the design, installation, maintenance, repair and approval of all chimneys, vents and connectors hereafter erected or altered in all building and structures.

1000.2 Other Standards. Unless otherwise specifically provided herein, conformity to the applicable requirements for chimney construction and vents contained in the mechanical code shall be deemed to meet the requirements of this code.

1000.3 Minor Repairs. Minor repairs for the purpose of maintenance and upkeep which do not increase the capacity of the heating apparatus or appliances, or which do not involve structural changes in the permanent chimney and vents of a building, may be made without a permit.

Section 1001.0 Plans and Specification

1001.1 General. The structural plans and specifications shall describe in sufficient detail, the location, size and construction of all chimneys, vents and ducts and their connections to boilers, furnaces, appliances and fireplaces. The thickness and character of all insulation materials, clearances from walls, partitions and ceilings and proximity of heating devices and equipment to wall openings and exitways shall be clearly shown and described.

1001.2 Appliances. All appliances required to be vented shall be connected to a vent or chimney, except as provided in Section 1006.3 and as provided in the standards for special venting arrangements.

Section 1002.0 Performance Test And Acceptance Criteria

1002.1 Tests. The building official may require a test or tests of any chimney or vent to insure fire safety and the removal of smoke and products of combustion.

1002.2 Acceptance Criteria. The system shall be accepted if the following conditions are fulfilled.

• There shall not be spillage at the draft hood when any one or combination of appliances connected to the system is in operation.

• Temperature on adjacent combustible surfaces shall not be raised more than limits acceptable to nationally recognized testing or inspection agencies.

• Condensation shall not be developed in a way that would cause deterioration of the vent or chimney drip from joints or bottom end of the vent or chimney.

• The draft reading taken at the place recommended in the installation instructions shall be within the range specified by the appliance manufacturer.

1002.2.1 Approved Installations. Factory-built chimneys and gas-vents which have been tested and listed by a nationally recognized testing or inspection agency shall be accepted as complying with the requirements of item 2 of Section 1002.2 when installed in accordance with the clearance specified in their listing.

Section 1003.0 Chimneys

Chimneys as used in this article shall be classified as:

- Factory-built chimneys.
- Masonry chimneys.
- Metal chimneys (smokestacks).

Section 1004.0 Appliances Requiring Chimneys

1004.1 General. All heating appliances except those appliances specifically exempted by the provisions of Section 1006.3 shall be connected to chimneys as specified in the chimney selection chart contained in the mechanical code.

Section 1005.0 Existing Buildings

1005.1 Raising Existing Chimneys. Whenever a building is hereafter erected, enlarged or increased in height so that a wall along an interior lot line, or within 3' thereof, extends above the top of an existing chimney or vent of an adjoining existing building, the owner of the building so erected, enlarged or increased in height shall carry up at his own expense, with the consent of the adjoining property owner, either independently, or in his own building, all chimneys connected to fuel burning appliances. Vents within 6' of any portion of the wall of such adjoining building shall be extended 2' above the roof or parapet of the adjoining building.

1005.2 Size Of Extended Chimneys. The construction of an extended chimney shall conform to the requirements of this article for new chimneys, but the internal area of such extension shall not be less than that of the existing chimney.

1005.3 Notice To Adjoining Owner. It shall be the duty of the owner of the building which is erected, enlarged or increased in height to notify in writing, and to secure the consent of, the owner of existing chimneys affected at least 10 days before starting such work.

1005.4 Existing Chimneys. An existing chimney, except one which does not endanger the fire safety of a building or structure and is acceptable to the building official, shall not be continued in use unless it conforms to all requirements of this article for new chimneys.

1005.5 Cleanouts And Maintenance. Whenever a new chimney is completed or an existing chimney is altered, it shall be cleaned and left smooth on the inside. If the chimney is constructed of masonry or tile, the interior mortar joints must be left smooth and flush. Cleanouts or other approved devices shall be provided at the base of all chimneys to enable the passageways to be maintained and cleaned.

Section 1006.0 Vent Systems

1006.1 Listed Appliances. For the purpose of determining vent requirements, gas-fired and oil-fired appliances shall be classified as "listed" or "unlisted." A listed appliance is one that is shown in a list published by an accredited authoritative testing agency, qualified and equipped for experimental testing of such appliances, and maintaining an adequate perioic inspection of current production of listed models and whole listing states either that the appliance or accessory complies with nationally recognized safety requirements or has been tested and found safe for use in a specific manner. Compliance may be determined by the presence on the appliance or accessory of a label of the testing agency stating that the appliance or accessory complies with nationally recognized safety requirements. An unlisted appliance or accessory is one that is not shown on such a list or does not bear such a label. In cases where an applicable standard has not been developed for a given class of appliance or accessory, approval of the authority having jurisdiction should be obtained before the appliance or accessory is installed.

1006.2 Appliances Required To Be Vented. Appliances shall be connected to a listed venting system or provided with other means for exhausting the flue gases to the outside atmosphere in accordance with the venting system selection chart.

1006.3 Exemption. Connections to vent systems shall not be required for appliances of such size or character that the absence of such connection does not constitute a hazard to the fire safety of the building or its occupants. The following appliances are not required to be vented unless so required to their listing:

• Listed gas ranges.

• Built-in domestic cooking units listed and marked as unvented units.

• Listed hot plates and listed laundry stoves.

• Listed domestic clothes dryers.

• Listed gas refrigerators.

• Counter appliances.

• Space (room) heaters listed for unvented use, only upon prior approval by the building official.

• Specialized equipment of limited input such as laboratory burners or gas lights.

• Electric appliances.

When any or all of the appliances listed in items 5, 6 and 7 above are installed so that the aggregate input rating exceeds 30 British thermal units (BTUs) per hour per cubic foot of room or space in which they are installed, one or more of them shall be vent connected or provided with approved means for exhausting the vent gases to the outside atmosphere so that the aggregate input rating of the remaining unvented appliance does not exceed 30 BTU per hour per cubic foot of room or space in which they are installed. Where the room or space in which they are installed is directly connected to another room or space by a doorway, arch, or other opening of comparable size, which cannot be closed, the volume of such adjacent room or space may be included in the calculation.

1006.4 **Types Of Vents.** Vents of common type include: Type B vents, Type B-W vents, and Type L low temperature vents.

Section 1007.0 Fireplaces

1007.1 General. Fireplaces, barbecues, smoke chambers and fireplace chimneys shall be of solid masonry or reinforced concrete or other approved materials, and shall conform to requirements of this section.

1007.2 Construction. Structural walls of fireplaces shall be at least eight (8) ″ thick. Where a lining of low duty refractory brick (ASTM C64) or the equivalent at least 2″ thick laid in fire clay mortar (ASTM C105, medium duty), or the equivalent, or other approved lining is provided, the total thickness of back and sides, including the lining, shall be not less than 8″. Where such lining is not provided, the thickness of back and sides shall be not less than 12″. The firebox shall be 20″ in depth and will be permited to be open on all sides, provided all fireplace openings are located entirely within one room.

1007.3 Lining. The lining shall extend from the throat of the fireplace to a point at least 4″ above the top of the enclosing masonry walls.

1007.4 Clearance.

1007.4.1 Distance. The distance between fireplace and combustibles shall be at least 4″ and such combustibles shall not be placed within 6″ of the fireplace opening. Wood facings or trim normally placed around the fireplace opening may be permitted when conforming to the requirements of this section; however, such facing or trim shall be furred out from the fireplace wall at least 4″ and attached to noncombustible furring strips. The edges of such facings or trim shall be covered with a noncombustible material. Where the walls of the fireplace are 12″ thick, the facings or trim may be directly attached to the fireplace.

1007.4.2 Metal Hoods. Metal hoods used as part of a fireplace or barbecue shall be at least 18″ from combustible material unless approved for reduced clearances.

1007.5 Metal. Metal hoods used as a part of a fireplace or barbecue shall be at least No. 18B&S (0.0403″) Gauge sheet copper, No. 18 Galvanized Steel Gage (0.052″) galvanized steel or other equivalent corrosion-resistant ferrous metal with all seams and connections of smokeproof unsoldered construction. The hoods shall be sloped at an angle of 45° or less from the vertical and shall extend horizontally at least 6″ beyond the limits of the firebox.

1007.6 Metal Heat Circulators. Approved metal heat circulators may be installed in fireplaces, provided the thickness of the fireplace walls is not reduced.

1007.7 Smoke Chamber. All walls, including back walls, shall be at least 8″ in thickness.

1007.8 Areas of Flues, Throats And Dampers. The net cross-sectional area of the flue and of the throat between the firebox and the smoke chamber of a fireplace shall be at least that required in Article 7. When dampers are used, damper openings shall be at least, when fully opened, equal to the required flue area and shall be of No. 12 Galvanized Steel Gauge (0.018″) metal.

1007.9 Lintel. Masonry over the fireplace opening shall be supported by a noncombustible lintel.

1007.10 Hearth. Every fireplace shall be constructed with a hearth of brick, stone, tile or other noncombustible material. For fireplaces with an opening of less than 6 square feet, the hearth shall extend not less than 16″ in front and not less than 8″ on each side of the fireplace opening. For fireplaces with an opening of 6 square feet or more, the hearth shall extend no less than 20″ in front and not less than 12″ on each side of the fireplace opening. Such hearths shall be supported on trimmer arches of brick, stone, tile or concrete not less than 4″ thick or other equally strong and fireresistance rated materials. All combustible forms or centering shall be removed after completion of the supporting construction.

1007.11 Firestopping. Firestopping between chimneys and wooden construction shall meet the requirements specified in Section 919.0 and the mechanical code.

1007.12 Support. Fireplaces shall be supported on foundations designed in conformity with Section 726.0.

1007.13 Screens. Screens or other acceptable protection devices shall be provided for all fireplace openings.

1007.14 Other Type Fireplaces. Other fireplaces not conforming to the requirements of this section shall be subject to approval by the department prior to installation. Imitation fireplaces shall not be used for the burning of gas, solid or liquid fuel. Approved factory-built fireplaces may be installed and shall conform to the applicable portions of this code. Factory-built fireplaces shall bear the seal of a nationally recognized testing or inspection agency.

1007.15 Solid Wastes. Solid waste shall not be burned in a fireplace.

Section 1008.0 Incinerators

1008.1 Mechanical Code. Incinerators of all types shall be installed in accordance with the applicable provisions of the mechanical code.

Section 1009.0 Construction of Metal Ducts And Vents

1009.1 Mechanical Code. All metal vents, ducts and duct systems required under the provisions of Articles 10 and 11 for heating systems and equipment, and under the provisions of Articles 5 and 18 for ventilating and air-conditioning systems shall be constructed and installed in accordance with the requirements of the mechanical code.

1009.2 Construction Of Ducts. Ducts and plenums may be constructed of approved material constructed in accordance with the requirements of the mechanical code.

Non-metallic ducts shall be constructed and installed in accordance with their approval and the applicable standards. Aluminum ducts shall not be used in equipment rooms with fuel-fired equipment, encased in or under concrete slabs on grade, for kitchen or fume exhausts or in systems where air entering the duct is over 250°F.

Section 1010.0 Spark Arrestors

1010.1 Mechanical Code. All chimneys, stacks and flues, including incinerator stacks, which emit sparks shall be provided with a spark arrestor conforming to the requirements of the mechanical code.

Appendix K
List Of Manufacturers

**American Standard Company
—Majestic**   *pre-built fireplaces*
245 Erie St.
Huntington, IN 46750

Franklin Cast Products   *wood burning stoves*
1800 Post Road
17 Airport Pazza
Warwick, RI 02886

Frontier—J & J Enterprises   *wood burning stoves*
4065 W. 11th Ave.
Eugene, OR 97402

Martin Industries   *pre-built fireplaces*
P. O. Box 1527
Huntsville, AL 35807

Suscon Stacks   *pre-built chimneys*
42 Fort Road
Magnolia, MD 21101

Vega Industries, Inc.—Heatilator *pre-built fireplaces*
19409 W. Saunders St.
Mt. Pleasant, IA 52241

Wallace Murry Corporation   *pre-built chimneys*
Metalbestos Systems
1820 E. Fargo
P. O. Box 372
Nampa, ID 83651

Appendix L
Useful Tables

Table L-1. Brick and Block Courses.

Block No. of Courses	Brick No. of Courses	Height of Course	Block No. of Courses	Brick No. of Courses	Height of Course
	1	0' 2-5/8"		37	8' 2-5/8"
	2	0' 5-3/8"		38	8' 5-3/8"
1	3	0' 8"	13	39	8' 8"
	4	0' 10-5/8"		40	8' 10-5/8"
	5	1' 1-3/8"		41	9' 1-3/8"
2	6	1' 4"	14	42	9' 4"
	7	1' 6-5/8"		43	9' 6-5/8"
	8	1' 9-3/8"		44	9' 9-3/8"
3	9	2' 0"	15	45	10' 0"
	10	2' 2-5/8"		46	10' 2-5/8"
	11	2' 5-3/8"		47	10' 5-3/8"
4	12	2' 8"	16	48	10' 8"
	13	2' 10-5/8"		49	10' 10-3/8"
	14	3' 1-3/8"		50	11' 1-3/8"
5	15	3' 4"	17	51	11' 4"
	16	3' 6-5/8"		52	11' 6-5/8"
	17	3' 9-3/8"		53	11' 9-3/8"
6	18	4' 0"	18	54	12' 0"
	19	4' 2-5/8"		55	12' 2-5/8"
	20	4' 5-3/8"		56	12' 5-3/8"
7	21	4' 8"	19	57	12' 8"
	22	4' 10-5/8"		58	12' 10-5/8"
	23	5' 1-3/8"		59	13' 1-3/8"
8	24	5' 4"	20	60	13' 4"
	25	5' 6-5/8"		61	13' 6-5/8"
	26	5' 9-3/8"		62	13' 9-3/8"
9	27	6' 0"	21	63	14' 0"
	28	6' 2-5/8"		64	14' 2-5/8"
	29	6' 5-3/8"		65	14' 5-3/8"
10	30	6' 8"	22	66	14' 8"
	31	6' 10-5/8"		67	14' 10-5/8"
	32	7' 1-3/8"		68	15' 1-3/8"
11	33	7' 4"	23	69	15' 4"
	34	7' 6-5/8"		70	15' 6-5/3"
	35	7' 9-3/8"		71	15' 9-3/8"
12	36	8' 0"	24	72	16' 0"

Mortar Joint is 3/8"

Table L-2. Safe Loads in Pounds per Square Foot on Different Types of Soil.

Material	Safe Load Lbs. Sq. Ft
Soft, wet clay or soft clay and wet sand mixed	2,000
Sand and clay—Firm clay or wet sand	4,000
Dry solid clay or firm dry sand	5,000
Hard clay—Firm coarse sand—Gravel	8,000
Firm coarse sand and gravel mixed	12,000
Hard Pan	20,000

Table L-3. Decimal Equivalents.

Decimal Equivalents

Decimal of a Foot						Decimal of an Inch	
Fraction	Decimal	Fraction	Decimal	Fraction	Decimal	Fraction	Decimal
1/16	0.0052	4-1/16	0.3385	8-1/16	0.6719	1/64	0.015625
1/8	0.0104	4-1/8	0.3438	8-1/8	0.6771	1/32	0.03125
3/16	0.0156	4-3/16	0.3490	8-3/16	0.6823	3/64	0.046875
1/4	0.0208	4-1/4	0.3542	8-1/4	0.6875	1/16	0.0625
5/16	0.0260	4-5/16	0.3594	8-5/16	0.6927	5/64	0.078125
3/8	0.0313	4-3/8	0.3646	8-3/8	0.6979	3/32	0.09375
7/16	0.0365	4-7/16	0.3698	8-7/16	0.7031	7/64	0.109375
1/2	0.0417	4-1/2	0.3750	8-1/2	0.7083	1/8	0.125
9/16	0.0459	4-9/16	0.3802	8-9/16	0.7135	9/64	0.140625
5/8	0.0521	4-5/8	0.3854	8-5/8	0.7188	5/32	0.15625
11/16	0.0573	4-11/16	0.3906	8-11/16	0.7240	11/64	0.171875
3/4	0.0625	4-3/4	0.3958	8-3/4	0.7292	3/16	0.1875
13/16	0.0677	4-13/16	0.4010	8-13/16	0.7344	13/64	0.203125
7/8	0.0729	4-7/8	0.4063	8-7/8	0.7396	7/32	0.21875
15/16	0.0781	4-15/16	0.4115	8-15/16	0.7448	15/64	0.234375
1-	0.0833	5-	0.4167	9-	0.7500	1/4	0.250
1-1/16	0.0885	5-1/16	0.4219	9-1/16	0.7552	17/64	0.265625
1-1/8	0.0938	5-1/8	0.4271	9-1/8	0.7604	9/32	0.28125
1-3/16	0.0990	5-3/16	0.4323	9-3/16	0.7656	19/64	0.296875
1-1/4	0.1042	5-1/4	0.4375	9-1/4	0.7708	5/16	0.3125
1-5/16	0.1094	5-5/16	0.4427	9-5/16	0.7760	21/64	0.328125
1-3/8	0.1146	5-3/8	0.4479	9-3/8	0.7813	11/32	0.34375
1-7/16	0.1198	5-7/16	0.4531	9-7/16	0.7865	23/64	0.359375
1-1/2	0.1250	5-1/2	0.4583	9-1/2	0.7917	3/8	0.375
1-9/16	0.1302	5-9/16	0.4635	9-9/16	0.7969	25/64	0.390625
1-5/8	0.1354	5-5/8	0.4688	9-5/8	0.8021	13/32	0.40625
1-11/16	0.1406	5-11/16	0.4740	9-11/16	0.8073	27/64	0.421875
1-3/4	0.1458	5-3/4	0.4792	9-3/4	0.8125	7/16	0.4375

1-13/16	0.1510	5-13/16	0.4844	9-13/16	0.8177	29/64	0.453125
1-7/8	0.1563	5-7/8	0.4896	9-7/8	0.8229	15/32	0.46875
1-15/16	0.1615	5-15/16	0.4948	9-15/16	0.8281	31/64	0.484375
2-	0.1667	6-	0.5000	10-	0.8333	1/2	0.500
2-1/16	0.1719	6-1/16	0.5052	10-1/16	0.8385	33/64	0.515625
2-1/8	0.1771	6-1/8	0.5104	10-1/8	0.8438	17/32	0.53125
2-3/16	0.1823	6-3/16	0.5156	10-3/16	0.8490	35/64	0.546875
2-1/4	0.1875	6-1/4	0.5208	10-1/4	0.8542	9/16	0.5625
2-5/16	0.1927	6-5/16	0.5260	10-5/16	0.8594	37/64	0.578125
2-3/8	0.1979	6-3/8	0.5313	10-3/8	0.8646	19/32	0.59375
2-7/16	0.2031	6-7/16	0.5365	10-7/16	0.8698	39/64	0.609375
2-1/2	0.2083	6-1/2	0.5417	10-1/2	0.8750	5/8	0.625
2-9/16	0.2135	6-9/16	0.5469	10-9/16	0.8802	41/64	0.640625
2-5/8	0.2188	6-5/8	0.5521	10-5/8	0.8854	21/32	0.65625
2-11/16	0.2240	6-11/16	0.5573	10-11/16	0.8906	43/64	0.671875
2-3/4	0.2292	6-3/4	0.5625	10-3/4	0.8958	11/16	0.6875
2-13/16	0.2344	6-13/16	0.5677	10-13/16	0.9010	45/64	0.703125
2-7/8	0.2396	6-7/8	0.5729	10-7/8	0.9063	23/32	0.71875
2-15/16	0.2448	6-15/16	0.5781	10-15/16	0.9115	47/64	0.734375
3-	0.2500	7-	0.5833	11-	0.9167	3/4	0.750
3-1/16	0.2552	7-1/16	0.5885	11-1/16	0.9219	49/64	0.765625
3-1/8	0.2604	7-1/8	0.5938	11-1/8	0.9271	25/32	0.78125
3-3/16	0.2656	7-3/16	0.5980	11-3/16	0.9323	51/64	0.796875
3-1/4	0.2708	7-1/4	0.6047	11-1/4	0.9375	13/16	0.8125
3-5/16	0.2760	7-5/16	0.6094	11-5/16	0.9427	53/64	0.828125
3-3/8	0.2813	7-3/8	0.6146	11-3/8	0.9479	27/32	0.84375
3-7/16	0.2865	7-7/16	0.6198	11-7/16	0.9531	55/64	0.859375
3-1/2	0.2917	7-1/2	0.6250	11-1/2	0.9583	7/8	0.875
3-9/16	0.2969	7-9/16	0.6302	11-9/16	0.9635	57/64	0.890625
3-5/8	0.3021	7-5/8	0.6354	11-5/8	0.9688	29/32	0.90625
3-11/16	0.3073	7-11/16	0.6406	11-11/16	0.9740	59/64	0.921875
3-3/4	0.3125	7-3/4	0.6458	11-3/4	0.9792	15/16	0.9375
3-13/16	0.3177	7-13/16	0.6510	11-13/16	0.9844	61/64	0.953125
3-7/8	0.3229	7-7/8	0.6563	11-7/8	0.9896	31/32	0.96875
3-15/16	0.3281	7-15/16	0.6615	11-15/16	0.9948	63/64	0.984375
4-	0.3333	8-	0.6667	12-	1.0000	1-	1.000

Table L-4. Various Measures and Conversion Factors.

Unit	Current U.S. term (multiply factor)	Conversion Factor*	SI term (divide factor)	SI Symbol
Length	inch	25.400	millimeter	mm
	foot	0.305	meter	m
	yard	0.914	meter	m
	mile	1.609	kilometer	km
Area	square inch	645.2	square millimeter	mm²
	square foot	0.093	square meter	m²
	square yard	0.836	square meter	m²
	square mile	2.590	square kilometer	km²
	acre	0.405	hectare	ha
Mass (weight)**	ounce	28.350	gram	g
	pound	0.454	kilogram	kg
	ton (2000 pounds)	0.907	metric ton	t
Volume	fluid ounce	29.574	milliliter	ml
	pint	0.473	liter	L
	quart	0.946	liter	L
	gallon	3.785	liter	L
	cubic foot	0.028	cubic meter	m³
	cubic yard	0.765	cubic meter	m³
	barrel (petroleum)	0.159	cubic meter	m³
Force	pound force	4.448	newton	N
Pressure and Stress	psi (pounds per square inch)	6.895	kilopascal	kPa

356

Category	Unit	Factor	Metric unit	Abbrev.
	psf (pounds per square foot)	.048	kilopascal	kPa
	ton per square foot	95.760	kilopascal	kPa
Electric Current†	ampere	no conversion	ampere	A
Light	lumen	no conversion	lumen	lm
	candela	no conversion	candela	cd
	foot candle	10.76	lux	lx
Heat, work or energy	Foot pound	1.356	joule	J
	kilowatt hour	3.600	megajoule	MJ
	BTU	1.055	kilojoule	kJ
Power	foot pound per second	1.355	watt	W
	BTU/hour	0.293	watt	W
	horse power	0.746	kilowatt	kW
	tons (refrigeration)	3.517	kilowatt	kW
Heat factors	U value	5.679	metric U value	undecided
	K value	1.730	metric K value	undecided
Temperature	degree Fahrenheit	††	degree Celcius	C
	degree Fahrenheit	††	Kelvin	K

* conversion factors have been rounded to the third decimal place
** mass and weight are not synonymous.
† electric terms already in common use.
†† to convert °F to °C subtract 32 and multiply by 1.8; to convert °F to K add 459.67 and multiply by 1.8

357

Table L-5. Metric Measures and Equivalents.

LINEAR MEASURE

10 millimeters	=	1 centimeter
10 centimeters	=	1 decimeter
10 decimeters	=	1 meter
10 meters	=	1 decameter
10 decameters	=	1 hectometer
10 hectometers	=	1 kilometer

SQUARE MEASURE

100 sq. millimeters	=	1 sq. centimeter
100 sq. centimeters	=	1 sq. decimeter
100 sq. decimeters	=	1 sq. meter
100 sq. meters	=	1 sq. decameter
100 sq. decameters	=	1 sq. hectometer
100 sq. hectometers	=	1 sq. kilometer

CUBIC MEASURE

1000 cu. millimeters	=	1 cu. centimeter
1000 cu. centimeters	=	1 cu. decimeter
100 cu. decimeters	=	1 cu. meter

LIQUID MEASURE

10 milliliters	=	1 centiliter
10 centiliters	=	1 deciliter
10 deciliters	=	1 liter
10 liters	=	1 decaliter
10 decaliters	=	1 hectoliter
10 hectoliters	=	1 kiloliter

WEIGHTS

10 milligrams	=	1 centigram
10 centigrams	=	1 decigram
10 decigrams		1 gram
10 grams		1 decagram
10 decagrams	=	1 hectogram
10 hectograms	=	1 kilogram
100 kilograms	=	1 quintal
10 quintals	=	1 ton

LINEAR MEASURE

1 inch	=		= 2.54 centimeters
1 foot	=	12 inches	= 0.3048 meter
1 yard	=	3 feet	= 0.9144 meter
1 rod	=	5 1/2 yds. or 16 1/2 ft.	= 5.029 meters
1 furlong	=	40 rods	= 201.17 meters
1 mile (statute)	=	5280 ft. or 1760 yds.	= 1609.3 meters
1 league (land)	=	3 miles	= 4.83 kilometers

SQUARE MEASURE

1 sq. inch	=		= 6.452 sq. centimeters
1 sq. foot	=	144 sq. inches	= 292 sq. centimeters
1 sq. yard	=	9 sq. feet	= 0.8361 sq. meter
1 sq. rod	=	30 1/4 sq. yards	= 25.29 sq. meters
1 acre	=	43,560 sq. feet or 160 sq. yds.	= 0.4047 hectare
1 sq. mile	=	640 acres	= 259 hectares or 2.59 sq. kilometers

CUBIC MEASURE

1 cu. inch	=		= 16.387 cu. centimeters
1 cu. foot	=	1728 cu. inches	= 0.0283 cu. meter
1 cu. yard	=	27 cu. feet	= 0.7646 cu. meter

ANGULAR AND CIRCULAR MEASURE

1 minute	=	60 seconds
1 degree	=	60 minutes
1 right angle	=	90 degrees
1 straight angle	=	180 degrees
1 circle	=	360 degrees

Table L-6. Weights of Building Materials.

Material	Weight	Material	Weight
CONCRETE		**WOOD CONSTRUCTION (CONTINUED)**	
With stone reinforced	150 pcf	Ceiling, joist and plaster	10 psf
With stone plain	144 pcf	Ceiling, joist and ½″ gypsum board	7 psf
With cinders, reinforced	110 pcf	Ceiling, joist and acoustic tile	5 psf
Light concrete (Aerocrete)	65 pcf	Wood shingeds	3 psf
(Perlite)	45 pcf	Spanish tile	15 psf
(Vermiculite)	40 pcf	Copper sheet	2 psf
		Tar and gravel	6 psf
METAL AND PLASTER			
Masonry mortar	116 pcf	**STONE**	
Gypsum and sand plaster	112 pcf	Sandstone	147 pcf
		Slate	175 pcf
BRICK AND BLOCK MASONRY (INCLUDING MORTAR)		Limestone	165 pcf
4″ brick wall	35 psf	Granite	175 pcf
8″ brick wall	74 psf	Marble	165 pcf
8″ concrete block wall	100 psf		
12″ concrete block wall	150 psf	**GLASS**	
4″ brick veneer over 4″ concrete block	65 psf	1 4″ plate glass	3.28 psf
		1 8″ double strength	1.63 psf
WOOD CONSTRUCTION		1 8″ insulating glass with air space	3.25 psk
Frame wall, lath and plaster	20 psf	4″ block glass	20.00 psf
Frame wall, 1 2″ gypsum board	12 psf		
Floor, 1 2″ subfloor - 3 4″ finished	6 psf	**INSULATION**	
Floor, 1 2″ subfloor and ceramic tile	16 psf	Cork board 1″ thick	58 psf
Roof, joist and ½″ sheathing	3 psf	Rigid foam insulation 2″ thick	3 psf
Roof, 2″ plank and beam	5 psf	Blanket or bat 2″ thick	1 psf
Roof, built-up	7 psf		

Appendix M
Photo Collection

Fig. M-1. View of the patio.

Fig. M-2. Note the sidewalk leading to the patio.

Fig. M-3. The antique bell at the entrance.

Fig. M-4. Note the television antenna on the pole.

Fig. M-5. Note the skylight.

Fig. M-6. An interesting landscape.

Fig. M-7. An aerial view.

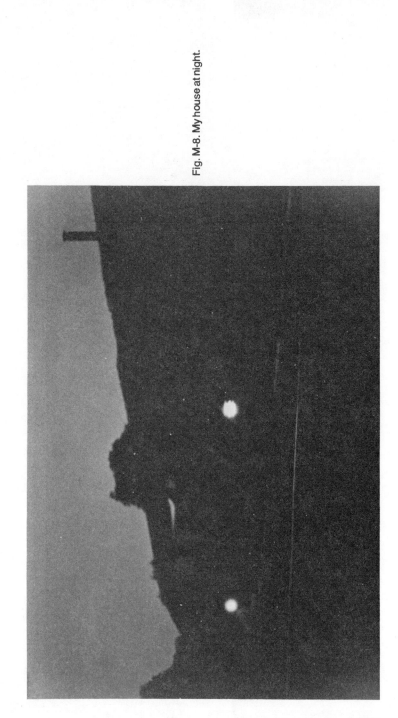

Fig. M-8. My house at night.

Fig. M-9. View of the driveway.

Fig. M-10. Most underground homes have skylights.

Fig. M-11. View of the driveway and adjacent area.

Fig. M-12. Shrubs help to enhance the appearance of a front yard.

Fig. M-13. A front view of my underground home.

Fig. M-14. Note the garage doors.

Fig. M-15. Note the lights.

Fig. M-16. My name and house number are displayed for all to see.

Fig. M-17. Another view of the patio.

Fig. M-18. I have to do quite a bit of grass cutting.

Fig. M-19. Another view of the skylight.

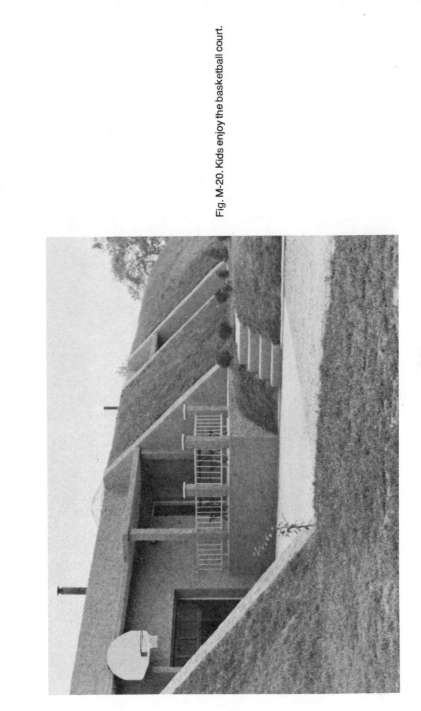

Fig. M-20. Kids enjoy the basketball court.

381

Fig. M-21. Another look at the old bell.

Fig. M-22. The driveway and the skylight.

Fig. M-23. A nice view of the sidewalk.

Fig. M-24. Note the picnic table.

Fig. M-25. My front lawn.

Fig. M-26. An entrance to my home.

Fig. M-27. Another view of the basketball court.

Fig. M-28. Shrubs, lights and the antique bell.

Fig. M-29. The skylight is in place over the garden.

Fig. M-30. The patio and basketball court.

Fig. M-31. Another view of the patio.

Fig. M-32. A nice view of the front yard.

Fig. M-33. Note the shrubs next to the sidewalk.

Fig. M-34. You're always welcome to visit my underground home.

Index

Index

Edited by Robert E. Ostrander